ARRL's
EXTRA

Q&A

Upgrade to an Extra Class Ham License!

By Ward Silver, NØAX

Contributing Editors
Mark Wilson, K1RO
Stephen Horan, AC5RI

Second Edition

Editorial Assistant
Maty Weinberg, KB1EIB

Production Staff
Michelle Bloom, WB1ENT, Production Supervisor
Jodi Morin, KA1JPA, Assistant Production Supervisor
Carol Michaud, KB1QAW, Production Assistant
Technical Illustrations: David Pingree, N1NAS
Cover Design: Sue Fagan, KB1OKW

ARRL The national association for
AMATEUR RADIO
225 Main Street, Newington, CT 06111-1494
www.arrl.org

This book may be used for Amateur Extra license exams given beginning July 1, 2008. *QST* and the ARRL Web site (**www. arrl.org**) will have news about any rules changes affecting the Extra class license or any of the material in this book.

We strive to produce books without errors. Sometimes mistakes do occur, however. When we become aware of problems in our books (other than obvious typographical errors), we post corrections on the ARRL Web site. If you think you have found an error, please check **www.arrl.org/notes** for corrections. If you don't find a correction there, please let us know, either using the Feedback Form at the back of this book or by sending e-mail to **pubsfdbk@arrl.org**.

Contents

Foreword

Now that you have decided to upgrade to the Extra class Amateur Radio license or are thinking about giving it a try, this book is designed to help. Welcome! Upgrading to Extra will complete your journey to the highest class of license Amateur Radio has to offer.

All of the information you need to pass the Element 4 Extra class written exam is here in the second edition of *ARRL's Extra Q & A*. Each and every one of the questions in the Extra class examination question pool is addressed with study material that explains the correct answer and provides background information. If you can answer the questions in this book, you can pass the written exam with confidence.

For more than 90 years, the ARRL has helped radio amateurs get the most out of their hobby by upgrading their licenses and advancing their operating and technical skills. The ARRL license preparation materials are the most complete package available to the amateur. *The ARRL Extra Class License Manual* is a detailed text reference that helps you understand the electronics theory, operating practices and FCC rules. This not only helps you pass the exam, but makes you more confident on the air and in your "shack."

If you'd like to take part in one of Amateur Radio's great traditions (and have a lot of fun, too), the ARRL provides complete training materials to learn the Morse code. While Morse proficiency is no longer required for Amateur licensing, it remains quite popular with amateurs for its power efficiency, simplicity and elegance.

This book and all of the ARRL technical material and operating aids are designed to help you do much more than just pass the written exam. There are books and supplies in the ARRL's "Radio Amateurs Library" that support almost any amateur operating practice today. Contact the Publication Sales Office at ARRL Headquarters to request the latest publications catalog or to place and order. (You can reach us by phone — 860-594-0200; by fax — 860-594-0303 and by email — **pubsales@arrl.org**. The complete publications catalog is on-line, too, at the ARRL's Web site — **www.arrl. org**. Look for the ARRL Product window, browse the latest news about Amateur Radio, and tap into the wealth of services provided by the ARRL to amateurs.

The second edition of *ARRL's Extra Q & A* is the product of cooperation between readers of the ARRL study materials, license class instructors, and the many ARRL staff members to bring the book to you. You can help make this book better by providing your own feedback. After you have passed your exam, write your suggestions, questions, and comments on the

Feedback Form at the back of the book, and send the form to us. Comments from readers are very important in making subsequent editions more effective and useful to readers.

Thanks for making the decision to upgrade — we hope to hear you on the air soon, using your new Extra class privileges and enjoying more of Amateur Radio. Good luck!

David Sumner, K1ZZ
Chief Executive Officer
Newington, Connecticut
March 2008

New Ham Desk	Prospective new amateurs call:
ARRL Headquarters	**800-32-NEW-HAM (800-326-3942)**
225 Main Street	You can also contact us via e-mail:
Newington, CT	**newham@arrl.org**
06111-1494	or check out **ARRLWeb**:
(860) 594-0200	**www.arrl.org/**

What is Amateur Radio?

Perhaps you've just picked up this book in the library or from a bookstore shelf and are wondering what this Amateur Radio business is all about. Maybe you have a friend or relative who is a "ham" and you're interested in becoming one, as well. In that case, a short explanation is in order.

Amateur Radio or "ham radio" is one of the longest-lived wireless activities. Amateur experimenters were operating right along with Marconi in the early part of the 20th century. They have helped advance the state-of-the-art in radio, television and dozens of other communications services since then, right up to the present day. There are more than half a million amateur radio operators or "hams" in the United States alone and several million more around the world!

Amateur Radio in the United States is a formal *communications service*, administered by the Federal Communications Commission or FCC. Created officially in its present form in 1934, the Amateur Service is intended to foster electronics and radio experimentation, provide emergency backup communications, encourage private citizens to train and practice operating, and even spread the goodwill of person-to-person contact over the airwaves.

Who Is a Ham and What Do Hams Do?

Anyone can be a ham — there are no age limits or physical requirements that prevent anyone from passing their license exam and getting on the air. Kids as young as 6 years old have passed the basic exam, and there are hams out there over the age of 100. You probably fall somewhere in the middle of that range.

Once you get on the air and start meeting other hams, you'll find a wide range of capabilities and interests. Of course, there are many technically skilled hams who work as engineers, scientists or technicians. But just as many don't have a deep technical background. You're just as likely to encounter writers, public safety personnel, students, farmers, truck drivers — anyone with an interest in personal communications over the radio.

The activities of Amateur Radio are incredibly varied. Amateurs who hold the Technician Class license — the usual first license for hams in the US — communicate primarily with local and regional amateurs using relay stations called *repeaters*. Known as "Techs," they sharp-

Ham radio, sun and fun. Peter Venlet, N8YEL, enjoys hilltop operating. Lightweight, portable rigs and small batteries give you many opportunities to pick your operating spot.

Bill Carter, KG4FXG, helps young Andrea Hartlage, KG4IUM, through her first Morse code contact.

Elaine Larson, KD6DUT, takes a turn at logging as Fred Martin, KI6YN, works the paddles during the Conejo Valley Amateur Radio Club's Field Day operation.

en their skills of operating while portable and mobile, often joining emergency communications teams. They may instead focus on the burgeoning wireless data networks assembled and used by hams around the world. Techs can make use of the growing number of Amateur Radio satellites, built and launched by hams along with the commercial "birds." Technicians transmit their own television signals, push the limits of signal propagation through the atmosphere and experiment with microwaves. Hams hold most of the world records for long-distance communication on microwave frequencies, in fact!

Bernardo Gonzalez, 6I2HWB, working into California from Mexico on a microwave frequency during the ARRL 10 GHz and Up Contest.

Hams who advance or *upgrade* to General and then to Extra Class are granted additional privileges with each step to use the frequencies usually associated with shortwave operation. This is the traditional Amateur Radio you probably encountered in movies or books. On these frequencies, signals can travel worldwide and so amateurs can make direct contact with foreign hams. No Internet, phone systems, or data networks are required. It's just you, your radio, and the ionosphere — the upper layers of the Earth's atmosphere!

Many hams use voice, Morse code, computer data modes and even image transmissions to communicate. All of these signals are mixed together where hams operate, making the experience of tuning a radio receiver through the crowded bands an interesting experience.

One thing common to all hams is that all of their operation is noncommercial, especially the volunteers who provide emergency communications. Hams pursue their hobby purely for personal enjoyment and to advance their skills, taking satisfaction from providing valuable services to their fellow citizens. This is especially valuable after natural disasters such as hurricanes and earthquakes when commercial systems are knocked out for a while. Amateur operators rush in to provide backup communication for hours, days, weeks or even months until the regular systems are restored. All this from a little study and a simple exam!

Want to Find Out More?

If you'd like to find out more about Amateur Radio in general, there is lots of information available on the Internet. A good place to start is on the American Radio Relay

Scott, KD5KZJ, making a contact on K5W/M, located aboard a refurbished 1927 rail car during a special "Troop Train" event on Memorial Day 2007.

League's (ARRL) ham radio introduction page at **www.hello-radio.org**. Books like *Ham Radio for Dummies* and *Getting Started With Ham Radio* will help you "fill in the blanks" as you learn more.

Along with books and Internet pages, there is no better way to learn about ham radio than to meet your local amateur operators. It is quite likely that no matter where you live in the United States, there is a ham radio club in your area — perhaps several! The ARRL provides a club lookup Web page at **www.arrl.org** where you can find a club just by entering your Zip code or state. Carrying on the tradition of mutual assistance, many clubs make helping newcomers to ham radio a part of their charter.

If this sounds like hams are confident that you'll find their activities interesting, you're right! Amateur Radio is much more than just talking on a radio, as you'll find out. It's an opportunity to dive into the fascinating world of radio communications, electronics, and computers as deeply as you wish to go. Welcome!

When to Expect
New Books

A Question Pool Committee (QPC) consisting of representatives from the various Volunteer Examiner Coordinators (VECs) prepares the license question pools. The QPC establishes a schedule for revising and implementing new Question Pools. The current Question Pool revision schedule is as follows:

Question Pool	Current Study Guides	Valid Through
Technician (Element 2)	The ARRL Ham Radio License Manual ARRL's Tech Q&A, 4th Ed.	June 30, 2010
General (Element 3)	The ARRL General Class License Manual, 6th Ed. ARRL's General Q&A, 3rd Ed.	June 30, 2011
Amateur Extra (Element 4)	The ARRL Extra Class License Manual, 9th Ed. ARRL's Extra Q&A, 2nd Ed.	June 30, 2012

As new question pools are released, ARRL will produce new study materials before the effective date of the new Pools. Until then, the current Question Pools will remain in use, and current ARRL study materials, including this book, will help you prepare for your exam.

As the new Question Pool schedules are confirmed, the information will be published in *QST* and on the ARRL Web site at **www.arrl.org**.

How to Use
This Book

To earn an Extra class Amateur Radio license, you must pass (or receive credit for) FCC Elements 2 (Technician Class), 3 (General Class) and 4 (Extra Class). This book is designed to help you prepare for and pass the Element 4 written exam. If you do not already have a General Amateur Radio license, you will need some additional study materials for the Element 3 (General) exam.

The Element 4 exam consists of 50 questions about Amateur Radio rules, theory and practice, as well as some electronics and radio theory. A passing grade is 74%, so you must answer 37 of the 50 questions correctly.

ARRL's Extra Q&A has 10 major sections that follow subelements E1 through E0 in the Extra class syllabus. The questions and multiple choice answers in this book are printed exactly as they were written by the Volunteer Examiner Coordinators' Question Pool Committee, and exactly as they will appear on your exam. (Be careful, though. The letter position of the answer choices may be scrambled, so you can't simply memorize an answer letter for each question.) In this book, the letter of the correct answer is printed in **boldface** type just before the explanation.

The ARRL also maintains a special Web page for Extra Class students at **www.arrl.org/eclm**. The useful and interesting on-line references listed there put you one click away from related and useful information.

If you are taking a licensing class, help your instructors by letting them know about areas in which you need help. They want you to learn as thoroughly and quickly as possible, so don't hold back with your questions. Similarly, if you find the material particularly clear or helpful, tell them that, too, so it can be used in the next class!

What We Assume About You

You don't have to be a technical guru or an expert operator to upgrade to Extra class! As you progress through the material, you'll build on the radio and electronics concepts you mastered for previous license exams. No advanced mathematics is introduced and if math gives you trouble, tutorials are listed at **www. arrl.org/eclm**. As with the General license, mastering rules and regulations will require learning some definitions and the finer points of regulations you may have already encountered. You should have a simple calculator, which you'll also be allowed to use during the license exam.

If you have some background in radio, perhaps as a technician or trained operator, you may be able to short-circuit some of the sections. It's common for technically-minded students to focus on the rules and regulations while students with an operating background tend to need the technical material more. Whichever you may be, be sure that you can answer the questions because they will certainly be on the test!

ARRL's Extra Q&A can be used either by an individual student, studying on his or her own, or as part of a licensing class taught by an instructor. If you're part of a class, the instructor will set the order in which the material is covered. The solo student can move at any pace and in any convenient order. You'll find that having a buddy to study with makes learning the material more fun as you help each other over the rough spots.

Don't hesitate to ask for help! If you can't find the answer in the book or at the Web site, email your question to the ARRL's New Ham Desk, **newham@arrl. org**. The ARRL's experts will answer directly or connect you with another ham that can answer your questions.

Online Exams

While you're studying and when you feel like you're ready for the actual exam you can get some good practice by taking one of the on-line Amateur Radio exams. These Web sites use the same Question Pool to construct an exam with the same number and variety of questions that you'll encounter on exam day. The exams are free and you can take them over and over again in complete privacy. Links to on-line exams can be found at **www.arrl.org/eclm**.

These exams are quite realistic and you get quick feedback about the questions you missed. When you find yourself passing the on-line exams by a comfortable margin, you'll be ready for the real thing! A note of caution, be sure that the questions used are from the current question pool.

Books to Help
You Learn

As you study the material on the licensing exam, you will have lots of other questions about the hows and whys of Amateur Radio. You may decide to experiment with new modes or techniques introduced for the exam. The following references, available from your local bookstore or the ARRL (**www.arrl.org/catalog**) will help "fill in the blanks" and give you a broader picture of the hobby:

• *ARRL Operating Manual.* With in-depth chapters on the most popular ham radio activities, this is your guide to what happens on the air. It even includes a healthy set of reference tables and maps.

• *Understanding Basic Electronics* by Larry Wolfgang, WR1B. Students who want more technical background about electronics will find this book helpful. It covers the fundamentals of electricity and electronics that are the foundation of all radio.

• *Basic Radio* by Joel Hallas, W1ZR. Students who are looking for in-depth knowledge about the technical elements of radio should take a look at this book. It covers the key building block of receivers, transmitters, antennas and propagation.

• *ARRL Handbook.* This is the grandfather of all Amateur Radio references and belongs on the shelf of all hams. Almost any topic you can think of in Amateur Radio technology is represented here.

• *ARRL Antenna Book.* After the radio itself, all radio depends on antennas. This book provides information on every common type of amateur antenna, feed lines and related topics, and construction tips and techniques.The solo student can move at any pace and in any convenient order. You'll find that having a buddy to study with makes learning the material more fun as you help each other over the rough spots.

Don't hesitate to ask for help! If you can't find the answer in the book or at the Web site, email your question to the ARRL's New Ham Desk, **newham@arrl. org**. The ARRL's experts will answer directly or connect you with another ham that can answer your questions.

Welcome to *ARRL's Extra Q&A*! Obtaining your Extra class license will enable you to make the most of your Amateur Radio experience. The additional HF band segments reserved for Extra class operators are the best for DXing and contesting, for example. Studying for the additional technical topics covered by the exam will help you get the most out of your current operating preferences.

This study guide will not only teach you the answers to the Extra class exam questions, but also provides explanations and supporting information. That way, you'll find it easier to learn the basic principles involved, and that will help you remember what you've learned. The book is full of useful facts and figures, so you'll want to keep it handy after you pass the test and are using your new privileges.

Do You Have a General License?

Most of this book's readers will have already earned their General class license, and some will have an Advanced license. Some readers may have been a ham for quite a while, and others may be new to the hobby. In either case, you're to be commended for making the effort to upgrade. We'll try to make it easy to pass your exam by teaching you the fundamentals and rationale behind each question and answer.

Reasons to Upgrade

If you're browsing through this book, trying to decide whether or not to upgrade, here are a few good reasons:

• *More fun.* The Extra class licensee has access to reserved sections of several of the HF bands. These *Extra class segments* are where most DX contacts are made and are considered prime territory. See **Table 1** for details of all of the frequencies reserved exclusively for Extra class licensees.

• *More communications options.* Your new understanding and skills will be valuable to your club, operating team or community.

• *New technical opportunities.* With your new understanding will come new ways of assembling and operating a station. Your improved technical understanding of electronics, radio and propagation will make you a more knowledgeable and skilled operator.

• *Volunteer examinations.* As an Extra class licensee, you'll be able to administer exams for any license class. Amateur Radio needs your help to make the volunteer licensing program work.

Not only does upgrading grant you complete Amateur Radio privileges, but by learning the material — perhaps even learning to operate using Morse code — your experiences will be much broader. You'll enjoy the hobby in ways that hams have pioneered and fostered for generations. The Extra's privileges are well worth your effort!

Table 1
Extra Class Band Segments

Band	Frequencies (MHz)
80 meters	3.500 – 3.525 and 3.600 – 3.700
40 meters	7.000 – 7.025
20 meters	14.000 – 14.025 and 14.150 – 14.175
15 meters	21.000 – 21.025 and 21.200 – 21.225

Table 2
Amateur License Class Examinations

License Class	Element Required	Number of Questions
Technician	2 (Written)	35 (passing is 26 correct)
General	3 (Written)	35 (passing is 26 correct)
Amateur Extra	4 (Written)	50 (passing is 37 correct)

Table 3
Exam Elements Needed to Qualify for an Amateur Extra License

Current License	Exam Requirements	Study Materials
None or Novice	Technician (Element 2)	The ARRL Ham Radio License Manual or ARRL's Tech Q&A
	General (Element 3)	The ARRL General Class License Manual or ARRL's General Q&A
	Extra (Element 4)	The ARRL Extra Class License Manual or ARRL's Extra Q&A (this book)
Technician*	General (Element 3)	The ARRL General Class License Manual or ARRL's General Q&A
	(Element 4)	The ARRL Extra Class License Manual or ARRL's Extra Q&A (this book)
General or Advanced	Extra (Element 4)	The ARRL Extra Class License Manual or ARRL's Extra Q&A (this book)

*Individuals who qualified for the Technician license before March 21, 1987, will be able to upgrade to General class by providing documentary proof at a VEC test session, paying an application fee and completing NCVEC Quick Form 605. No additional exam is required.

Extra Class Overview

There are three classes of license being granted today: Technician, General and Amateur Extra. Each grants the licensee more and more privileges, meaning access to frequencies and modes. Extra class licensees have all amateur privileges. **Table 2** shows the elements for each of the amateur licenses as of early 2008.

As shown in Table 2, to qualify for an Extra class license, you must have passed Elements 2 (Technician), 3 (General) and 4 (Extra). If you hold a General or Advanced license, you are credited with Elements 2 and 3, so you don't have to take them again. If you have a Technician license issued before March 21, 1987, you can upgrade to General simply by going to a test session with proof of being licensed before that date. **Table 3** shows the variations of licenses and element credit.

The 50-question multiple-choice test for Element 4 is more comprehensive than the Element 3 General exam because as a more experienced ham, your wider knowledge will allow you to experiment with, modify, and build more types of equipment and antennas. An Extra class amateur can use any mode, frequency or technique available to amateurs.

Morse Code

Although you no longer need to learn Morse code for any license exam, Morse code or "CW" has been part of the rich amateur tradition for 100 years and many hams still use it extensively. There are solid reasons for it to be used, too! It's easy to build Morse code transmitters and receivers. There is no more power-efficient mode of communication that is copied by the human ear. The extensive set of prosigns and signals allows amateurs to communicate a great deal of information even if they don't share a common language. Morse is likely to remain part of the amateur experience for a long time to come.

If you are interested in learning Morse code, the ARRL has a complete set of resources listed on its Web page at **www.arrl.org/FandES/ead/learncw**. Computer software and on-the-air code practice sessions are available for personal training and practice. Organizations such as FISTS (**www.fists.org**) — an operator's style of sending is referred to as his or her "fist" — help hams learn Morse code and will even help you find a "code buddy" to share the learning with you.

Want More Information?

Looking for more information about Extra class instruction in your area? Are you ready to take the Extra class exam? Do you need a list of ham radio clubs, instructors or examiners in your local area? The following Web pages are very helpful in finding the local resources you need to successfully pass your Extra exam:

www.arrl.org — the ARRL's home page, it features news and links to other ARRL resources

www.arrl.org/eclm — the Web site that supports this book

www.arrl.org/FandES/field/club/clubsearch.phtml — a search page to find ARRL-Affiliated clubs

www.arrl.org/arrlvec/examsearch.phtml — the ARRL/VEC exam session search page

www.arrl.org/tis — the ARRL's Technical Information Service is an excellent resource

www.ac6v.org — a Web site that compiles links to hundreds of ham radio Web pages

David Joe Williams AJ5W and Harry Mueller, KC5TRB, inspect a payload after a balloon experiment tracked by the Amateur Packet Reporting System (APRS).

It took plenty of aloha to make the first 13 cm (2304 MHz) contact between Hawaii (KH6) and the mainland US using EME, but (l-r) KH6ND, KH7U, KØYW, WH6GS and AH6NF (host KH6YY not pictured) set up a temporary station to do the job.

The Upgrade Trail

As you begin your studies remember that you've already overcome two big hurdles — taking and passing the Technician and General license exams! The questions may be more numerous and challenging for the Extra class exam, but you already know all about the testing procedure and the basics of ham radio. You can approach the process of upgrading with confidence!

Refine Your Knowledge

The Extra class exam mostly deals with the finer points of operating, the FCC's rules, and the technical details of radio. We'll cover more advanced modes and signals, too. The goal is to help you "fill in the blanks" in your ham radio knowledge. Here are some examples of topics that you'll be studying:

- Impedance and resonance
- Image modes — ATV and SSTV
- How digital protocols work
- Long-path, grey line and transequatorial propagation
- Intermodulation and receiver performance
- Practical radio circuits
- More types of antennas

Not every ham uses every mode and frequency, of course. By mastering a wider range of skills and knowledge, hams can get more out of everyday operating. Extra class hams can take the lead in planning and assembling stations and operating teams. Having deeper knowledge of radio will lead you to a greater appreciation of the magic of radio!

Testing Process

When you're ready, you'll need to find a test session. If you're in a licensing class, the instructor will help you find and register for a session. Otherwise, you can find a test session by using the ARRL's Web page for finding exams, **www.arrl.org/arrlvec/examsearch.phtml**. If you can register for the test session in advance, do so. Other sessions, such as those at hamfests or conventions, are available to anyone that shows up (called *walk-ins*). You may have to wait for an available space though, so go early!

This sample NCVEC Quick Form 605 shows how your form will look after you have completed your upgrade to Extra.

Quick-Form Application for Authorization in the Ship, Aircraft, Amateur, Restricted and Commercial Operator, and General Mobile Radio Services

Approved by OMB
3060 - 0850
See instructions for
public burden estimate

1) Radio Service Code:	HA

Application Purpose (Select only one) (MD)

2) NE – New	RO – Renewal Only	WD – Withdrawal of Application
MD – Modification	RM – Renewal / Modification	DU – Duplicate License
AM – Amendment	CA – Cancellation of License	AU – Administrative Update

3) If this request if for Developmental License or STA (Special Temporary Authorization) enter the appropriate code and attach the required exhibit as described in the instructions. Otherwise enter 'N' (Not Applicable).	(N) D S N/A
4) If this request is for an Amendment or Withdrawal of Application, enter the file number of the pending application currently on file with the FCC.	File Number
5) If this request is for a Modification, Renewal Only, Renewal / Modification, Cancellation of License, Duplicate License, or Administrative Update, enter the call sign (serial number for Commercial Operator) of the existing FCC license. If this is a request for consolidation of DO & DM Operator Licenses, enter serial number of DO. Also, if filing for a ship exemption, you must provide call sign.	Call Sign/Serial # AB1FM
6) If this request is for a New, Amendment, Renewal Only, or Renewal Modification, enter the requested expiration date of the authorization (this item is optional).	MM DD
7) Does this filing request a Waiver of the Commission's rules? If 'Y', attach the required showing as described in the instructions.	(N) Yes No
8) Are attachments (other than associated schedules) being filed with this application?	(N) Yes No

Applicant/Licensee Information

9) FCC Registration Number (FRN):   0012345678

10) Applicant/Licensee legal entity type: (Select One)

☒ Individual   ☐ Corporation   ☐ Unincorporated Association   ☐ Trust   ☐ Government Entity
☐ Consortium   ☐ General Partnership   ☐ Limited Liability Company   ☐ Limited Liability Partnership
☐ Limited Partnership   ☐ Other (Description of Legal Entity)

11) First Name (if individual): MARIA	MI: A	Last Name: SOMMA	Suffix:

12) Entity Name (if other than individual):

13) If the licensee name is being updated, is the update a result from the sale (or transfer of control) of the license(s) to another party and for which proper Commission approval has not been received or proper notification not provided?   (   ) Yes   No

14) Attention To:

15) P.O. Box:	And/Or	16) Street Address: 225 MAIN ST.	
17) City: NEWINGTON	18) State: CT	19) Zip Code/Postal Code: 06066	20) Country:

21) Telephone Number:	22) FAX Number:

23) E-Mail Address:

Ship Applicants/Licensees Only

24) Enter new name of vessel:_____

Aircraft Applicants/Licensees Only

25) Enter the new FAA Registration Number (the N-number):_____
  NOTE: Do not enter the leading "N".

Portions of FCC Form 605 showing the sections you would complete for a modification of your license, such as a change of address.

As for all amateur exams, the Extra class exam is administered by Volunteer Examiners (VEs). All VEs are certified by a Volunteer Examiner Coordinator (VEC) such as the ARRL/VEC. This organization trains and certifies VEs and processes the FCC paperwork for their test sessions.

Bring your current license *original* and a photocopy (to send with the application). You'll need two forms of identification including at least one photo ID, such as a driver's license, passport or employer's identity card. Know your Social Security

Fee Status

26) Is the applicant/licensee exempt from FCC application Fees?	(N) Yes No
27) Is the applicant/licensee exempt from FCC regulatory Fees?	(N) Yes No

General Certification Statements

1) The applicant/licensee waives any claim to the use of any particular frequency or of the electromagnetic spectrum as against the regulatory power of the United States because of the previous use of the same, whether by license or otherwise, and requests an authorization in accordance with this application.

2) The applicant/licensee certifies that all statements made in this application and in the exhibits, attachments, or documents incorporated by reference are material, are part of this application, and are true, complete, correct, and made in good faith.

3) Neither the applicant/licensee nor any member thereof is a foreign government or a representative thereof.

4) The applicant/licensee certifies that neither the applicant/licensee nor any other party to the application is subject to a denial of Federal benefits pursuant to Section 5301 of the Anti-Drug Abuse Act of 1988, 21 U.S.C. § 862, because of a conviction for possession or distribution of a controlled substance. **This certification does not apply to applications filed in services exempted under Section 1.2002(c) of the rules, 47 CFR § 1.2002(c).** See Section 1.2002(b) of the rules, 47 CFR § 1.2002(b), for the definition of "party to the application" as used in this certification.

5) Amateur or GMRS applicant/licensee certifies that the construction of the station would NOT be an action which is likely to have a significant environmental effect (see the Commission's rules 47 CFR Sections 1.1301-1.1319 and Section 97.13(a) rules (available at web site http://wireless.fcc.gov/rules.html).

6) Amateur applicant/licensee certifies that they have READ and WILL COMPLY WITH Section 97.13(c) of the Commission's rules (available at web site http://wireless.fcc.gov/rules.html) regarding RADIOFREQUENCY (RF) RADIATION SAFETY and the amateur service section of OST/OET Bulletin Number 65 (available at web site http://www.fcc.gov/oet/info/documents/bulletins/).

Certification Statements For GMRS Applicants/Licensees

1) Applicant/Licensee certifies that he or she is claiming eligibility under Rule Section 95.5 of the Commission's rules.

2) Applicant/Licensee certifies that he or she is at least 18 years of age.

3) Applicant/Licensee certifies that he or she will comply with the requirement that use of frequencies 462.650, 467.650, 462.700 and 467.700 MHz is not permitted near the Canadian border North of Line A and East of Line C. These frequencies are used throughout Canada and harmful interference is anticipated.

4) Non-Individual applicants/licensees certify that they have NOT changed frequency or channel pairs, type of emission, antenna height, location of fixed transmitters, number of mobile units, area of mobile operation, or increase in power.

Certification Statements for Ship Applicants/Licensees (Including Ship Exemptions)

1) Applicant/Licensee certifies that they are the owner or operator of the vessel, a subsidiary communications corporation of the owner or operator of the vessel, a state or local government subdivision, or an agency of the US Government subject to Section 301 of the Communications Act.

2) This application is filed with the understanding that any action by the Commission thereon shall be limited to the voyage(s) described herein, and that apart from the provisions of the specific law from which the applicant/licensee requests an exemption, the vessel is in full compliance with all applicable statues, international agreements and regulations.

Signature

28) Typed or Printed Name of Party Authorized to Sign

First Name: MARIA	MI: A	Last Name: SOMMA	Suffix:

29) Title:

Signature: *Maria Somma*	30) Date: 04/01/06

Failure to Sign This Application May Result in Dismissal Of The Application And Forfeiture Of Any Fees Paid

WILLFUL FALSE STATEMENTS MADE ON THIS FORM OR ANY ATTACHMENTS ARE PUNISHABLE BY FINE AND/OR IMPRISONMENT (U.S. Code, Title 18, Section 1001) AND / OR REVOCATION OF ANY STATION LICENSE OR CONSTRUCTION PERMIT (U.S. Code, Title 47, Section 312(a)(1)), AND / OR FORFEITURE (U.S. Code, Title 47, Section 503).

FCC 605 – Main Form
July 2005 - Page 2

Number (SSN) or FCC Federal Registration Number (FRN). You can bring pencils or pens, blank scratch paper and a calculator, but any kind of computer or on-line device is prohibited.

The FCC allows Volunteer Examiners to use a range of procedures to accommodate applicants with various disabilities. If this applies to you, you'll still have to pass the test, but special exam procedures can be applied. Contact your local VE team or the Volunteer Examiner Coordinator (VEC) responsible for the test session you'll be attending. Contact the ARRL/VEC Office at 225 Main St, Newington CT 06111-1494 or by phone at 860-594-0200.

Once you're signed in, you'll need to fill out a copy of the National Conference of Volunteer Examiner Coordinator's (NCVEC) Quick Form 605. This is an application for a new or upgraded license. It is used only at test sessions and for a VEC to process a license renewal or a license change. *Do not* use an NCVEC Quick Form 605 for any kind of application directly to the FCC — it will be rejected. Use a regular FCC Form 605. After filling out the form, pay the current test fee and get ready.

The Exam

To complete the Extra exam takes less than an hour. You will be given a question booklet and an answer sheet. Be sure to read the instructions, fill in all the necessary information and sign your name wherever required. Check to be sure your booklet has all the questions and be sure to mark the answer in the correct space for each question.

You don't have to answer the questions in order — skip the hard ones and go back to them. If you read the answers carefully, you'll probably find that you can eliminate one or more "distractors." Of the remaining answers, only one will be the best. If you can't decide which is the correct answer, go ahead and guess. There is no penalty for an incorrect guess. When you're done, go back and check your answers and double-check your arithmetic — there's no rush!

Once you've answered all 50 questions, the VEs will grade and verify your test results. Assuming you've passed (congratulations!) you'll fill out a *Certificate of Successful Completion of Examination* (CSCE). The exam organizers will submit your results to the FCC while you keep one copy of the CSCE as evidence that you've passed your Amateur Extra test.

If you are licensed and already have a call sign, you can begin using your new privileges immediately. When you give your call sign, append "/AE" (on CW or digital modes) or "temporary AE" (on phone). As soon as your name and call sign appear in the FCC's database of licensees, typically a week to 10 days later, you can stop adding the suffix. The CSCE is good for 365 days in case you need to retake an exam before receiving your paper license.

If you don't pass, don't be discouraged! You might be able to take another version of the test right then and there if the session organizers can accommodate you. Even if you decide to try again later, you now know just how the test session feels — you'll be more relaxed and ready next time. The ham bands are full of hams who took their Extra test more than once before passing. You'll be in good company!

FCC and ARRL/VEC Licensing Resources

After you pass your exam, the examiners will file all of the necessary paperwork so that your license will be granted by the Federal Communications Commission (FCC). In a few days, you will be able see your new call sign in the FCC's database via the ARRL's Web site and later, you'll receive a paper license by mail.

When you passed an earlier exam, you may have applied for your FCC Federal Registration Number (FRN). This allows you to access the information for any FCC licenses you may have and to request modifications to them. These functions are available via the FCC's Universal Licensing System Web site (**wireless.fcc.gov/**

The ARRL VEC processes tens of thousands of license applications annually.

uls) and complete instructions for using the site are available at **www.arrl.org/fcc/uls-qa.html**.

The ARRL/VEC can also process license renewals and modifications for you as described at **www.arrl.org/arrlvec/renewals.html**.

Time to Get Started

By following these instructions and carefully studying the material in this book, soon you'll be operating in the Extra class segments with the rest of the Extras! Each of us at the ARRL Headquarters and every ARRL member looks forward to the day when you join the fun. '73' (best regards) and good luck!

Table 4
Extra Class (Element 4) Syllabus
SUBELEMENT E1 — COMMISSION'S RULES

[6 Exam Questions — 6 Groups]

E1A Operating Standards: frequency privileges for Extra class amateurs; emission standards; automatic message forwarding; frequency sharing; FCC license actions; stations aboard ships or aircraft

E1B Station restrictions and special operations: restrictions on station location; general operating restrictions, spurious emissions, control operator reimbursement; antenna structure restrictions; RACES operations

E1C Station control: definitions and restrictions pertaining to local, automatic and remote control operation; control operator responsibilities for remote and automatically controlled stations

E1D Amateur Satellite service: definitions and purpose; license requirements for space stations; available frequencies and bands; telecommand and telemetry operations; restrictions, and special provisions; notification requirements

E1E Volunteer examiner program: definitions, qualifications, preparation and administration of exams; accreditation; question pools; documentation requirements

E1F Miscellaneous rules: external RF power amplifiers; Line A; national quiet zone; business communications; compensated communications; spread spectrum; auxiliary stations; reciprocal operating privileges; IARP and CEPT licenses; third party communications with foreign countries; special temporary authority

SUBELEMENT E2 — OPERATING PRACTICES AND PROCEDURES

[5 Exam Questions — 5 Groups]

E2A Amateur radio in space: amateur satellites; orbital mechanics; frequencies and modes; satellite hardware; satellite operations

E2B Television practices: fast scan television standards and techniques; slow scan television standards and techniques

E2C Operating methods, part 1: contest and DX operating; spread-spectrum transmissions; automatic HF forwarding; selecting an operating frequency

E2D Operating methods, part 2: VHF and UHF digital modes; packet clusters; Automatic Position Reporting System (APRS)

E2E Operating methods, part 3: operating HF digital modes; error correction

SUBELEMENT E3 — RADIO WAVE PROPAGATION

[3 Exam Questions — 3 Groups]

E3A Propagation and technique, part 1: Earth-Moon-Earth communications; meteor scatter

E3B Propagation and technique, part 2: transequatorial; long path; gray line; multi-path propagation

E3C   Propagation and technique, part 3: Auroral propagation; selective fading; radio-path horizon; take-off angle over flat or sloping terrain; earth effects on propagation; less common propagation modes

SUBELEMENT E4 — AMATEUR RADIO TECHNOLOGY AND MEASUREMENTS

[5 Exam Questions — 5 Groups]

E4A   Test equipment: analog and digital instruments; spectrum and network analyzers, antenna analyzers; oscilloscopes; testing transistors; RF measurements

E4B   Measurement technique and limitations: instrument accuracy and performance limitations; probes; techniques to minimize errors; measurement of "Q"; instrument calibration

E4C   Receiver performance characteristics, part 1: phase noise, capture effect, noise floor, image rejection, MDS, signal-to-noise-ratio; selectivity

E4D   Receiver performance characteristics, part 2: blocking dynamic range, intermodulation and cross-modulation interference; 3rd order intercept; desensitization; preselection

E4E   Noise suppression: system noise; electrical appliance noise; line noise; locating noise sources; DSP noise reduction; noise blankers

SUBELEMENT E5 — ELECTRICAL PRINCIPLES

[4 Exam Questions — 4 Groups]

E5A   Resonance and Q: characteristics of resonant circuits: series and parallel resonance; Q; half-power bandwidth; phase relationships in reactive circuits

E5B   Time constants and phase relationships: R/L/C time constants: definition; time constants in RL and RC circuits; phase angle between voltage and current; phase angles of series and parallel circuits

E5C   Impedance plots and coordinate systems: plotting impedances in polar coordinates; rectangular coordinates

E5D   AC and RF energy in real circuits: skin effect; electrostatic and electromagnetic fields; reactive power; power factor; coordinate systems

SUBELEMENT E6 — CIRCUIT COMPONENTS

[6 Exam Questions — 6 Groups]

E6A   Semiconductor materials and devices: semiconductor materials (germanium, silicon, P-type, N-type); transistor types: NPN, PNP, junction, power; field-effect transistors: enhancement mode; depletion mode; MOS; CMOS; N-channel; P-channel

E6B   Semiconductor diodes

E6C   Integrated circuits: TTL digital integrated circuits; CMOS digital integrated circuits; gates

E6D Optical devices and toroids: vidicon and cathode-ray tube devices; charge-coupled devices (CCDs); liquid crystal displays (LCDs); toroids: permeability, core material, selecting, winding

E6E Piezoelectric crystals and MMICS: quartz crystals (as used in oscillators and filters); monolithic amplifiers (MMICs)

E6F Optical components and power systems: photoconductive principles and effects, photovoltaic systems, optical couplers, optical sensors, and optoisolators

SUBELEMENT E7 — PRACTICAL CIRCUITS

[8 Exam Questions — 8 Groups]

E7A Digital circuits: digital circuit principles and logic circuits: classes of logic elements; positive and negative logic; frequency dividers; truth tables

E7B Amplifiers: Class of operation; vacuum tube and solid-state circuits; distortion and intermodulation; spurious and parasitic suppression; microwave amplifiers

E7C Filters and matching networks: filters and impedance matching networks: types of networks; types of filters; filter applications; filter characteristics; impedance matching; DSP filtering

E7D Power supplies and voltage regulators

E7E Modulation and demodulation: reactance, phase and balanced modulators; detectors; mixer stages; DSP modulation and demodulation; software defined radio systems

E7F Frequency markers and counters: frequency divider circuits; frequency marker generators; frequency counters

E7G Active filters and op-amps: active audio filters; characteristics; basic circuit design; operational amplifiers

E7H Oscillators and signal sources: types of oscillators; synthesizers and phase-locked loops; direct digital synthesizers

SUBELEMENT E8 — SIGNALS AND EMISSIONS

[4 Exam Questions — 4 Groups]

E8A AC waveforms: sine, square, sawtooth and irregular waveforms; AC measurements; average and PEP of RF signals; pulse and digital signal waveforms

E8B Modulation and demodulation: modulation methods; modulation index and deviation ratio; pulse modulation; frequency and time division multiplexing

E8C Digital signals: digital communications modes; CW; information rate vs. bandwidth; spread-spectrum communications; modulation methods

E8D Waves, measurements, and RF grounding: peak-to-peak values, polarization; RF grounding

SUBELEMENT E9 — ANTENNAS AND TRANSMISSION LINES

[8 Exam Questions — 8 Groups]

E9A Isotropic and gain antennas: definition; used as a standard for comparison; radiation pattern; basic antenna parameters: radiation resistance and reactance, gain, beamwidth, efficiency

E9B Antenna patterns: E and H plane patterns; gain as a function of pattern; antenna design (computer modeling of antennas); Yagi antennas

E9C Wire and phased vertical antennas: beverage antennas; terminated and resonant rhombic antennas; elevation above real ground; ground effects as related to polarization; take-off angles

E9D Directional antennas: gain; satellite antennas; antenna beamwidth; losses; SWR bandwidth; antenna efficiency; shortened and mobile antennas; grounding

E9E Matching: matching antennas to feed lines; power dividers

E9F Transmission lines: characteristics of open and shorted feed lines: ⅛ wavelength; ¼ wavelength; ½ wavelength; feed lines: coax versus open-wire; velocity factor; electrical length; transformation characteristics of line terminated in impedance not equal to characteristic impedance

E9G The Smith chart

E9H Effective radiated power; system gains and losses; radio direction finding antennas

SUBELEMENT E0 — Safety

[1 Exam Question — 1 Group]

E0A Safety: amateur radio safety practices; RF radiation hazards; hazardous materials

Commission's Rules

The Extra Class (Element 4) written examination consists of 50 questions taken from the Extra Class examination pool. This pool is prepared by the Volunteer Examiner Coordinators' Question Pool Committee. A certain number of questions are taken from each of the 10 subelements (numbered E1 through E0).

There will be six examination questions over the six groups of questions covering the Commission's Rules. The question groups are labeled E1A through E1F.

The correct answer (A, B, C or D) is given in bold following the question and the possible responses at the beginning of an explanation section. This convention will be used throughout this book.

After many of the explanations in this subelement, you will see a reference to Part 97 of the FCC rules set inside square brackets, like [97.301(b)]. This tells you where to look for the exact wording in the Rules as they relate to that question. For a complete copy of Part 97, along with simple explanations of the Rules governing Amateur Radio, see the FCC Rules online at **www.arrl.org/FandES/field/regulations/rules-regs.html**. In addition to Part 97, you'll find references to other parts of the FCC rules. These include Part 1 and Part 17.

E1A   Operating Standards: frequency privileges for Extra Class amateurs; emission standards; automatic message forwarding; frequency sharing; FCC license actions; stations aboard ships or aircraft

E1A01   When using a transceiver that displays the carrier frequency of phone signals, which of the following displayed frequencies will result in a normal USB emission being within the band?

A. The exact upper band edge
B. 300 Hz below the upper band edge
C. 1 kHz below the upper band edge
D. 3 kHz below the upper band edge

D   The components that make up the sideband of a USB signal are higher than the carrier frequency. Since an amateur SSB signal generally has a bandwidth of about 3 kHz, to be sure the sideband components of a USB signal are within the amateur band, the carrier frequency should be 3 kHz below the band edge. [97.301, 97.305]

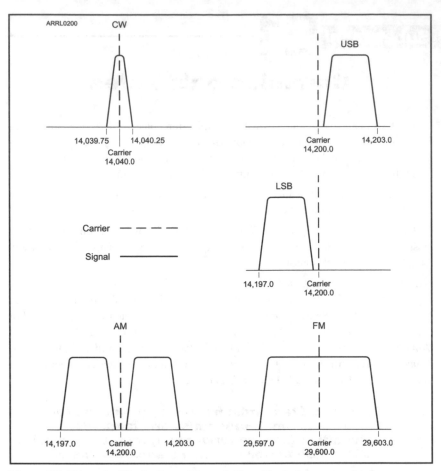

Figure E1A01 — An illustration of the relationship between carrier frequency and the actual signal energy. Most rigs are configured to show the carrier frequency of a signal.

E1A02 When using a transceiver that displays the carrier frequency of phone signals, which of the following displayed frequencies will result in a normal LSB emission being within the band?

A. The exact lower band edge
B. 300 Hz above the lower band edge
C. 1 kHz above the lower band edge
D. 3 kHz above the lower band edge

D The components that make up the sideband of an LSB signal are lower than the carrier frequency. Since an amateur SSB signal generally has a bandwidth of about 3 kHz, to be sure the sideband components of an LSB signal are within the amateur band, the carrier frequency should be 3 kHz above the band edge. [97.301, 97.305]

E1A03 With your transceiver displaying the carrier frequency of phone signals, you hear a DX station's CQ on 14.349 MHz USB. Is it legal to return the call using upper sideband on the same frequency?

A. Yes, because the DX station initiated the contact
B. Yes, because the displayed frequency is within the 20 meter band
C. No, my sidebands will extend beyond the band edge
D. No, USA stations are not permitted to use phone emissions above 14.340 MHz

C   See the discussion for E1A01. [97.301, 97.305]

E1A04 With your transceiver displaying the carrier frequency of phone signals, you hear a DX station's CQ on 3.601 MHz LSB. Is it legal to return the call using lower sideband on the same frequency?

A. Yes, because the DX station initiated the contact
B. Yes, because the displayed frequency is within the 75 meter phone band segment
C. No, my sidebands will extend beyond the edge of the phone band segment
D. No, USA stations are not permitted to use phone emissions below 3.610 MHz

C   See the discussion for E1A02. [97.301, 97.305]

E1A05 Which is the only amateur band that does not permit the transmission of phone or image emissions?

A. 160 meters
B. 60 meters
C. 30 meters
D. 17 meters

C   30 meters is restricted to CW and data emissions as shown in the table of Operating Privileges. [97.305]

US Amateur Bands

Effective February 23, 2007

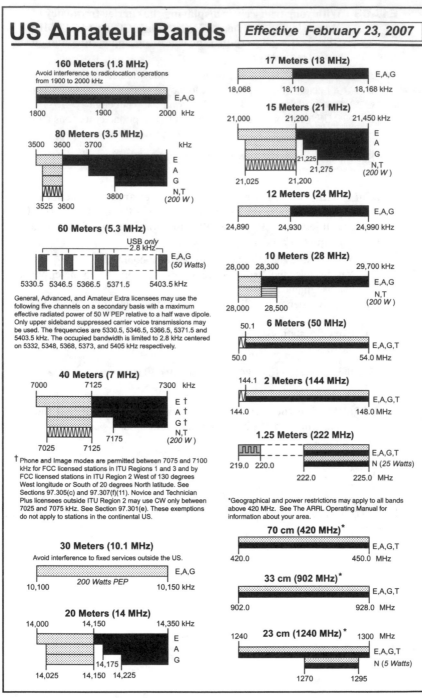

Figure E1A05 — Amateur operating privileges

E1A06 What is the maximum power output permitted on the 60 meter band?

A. 50 watts PEP effective radiated power relative to an isotropic radiator
B. 50 watts PEP effective radiated power relative to a dipole
C. 100 watts PEP effective radiated power relative to an isotropic radiator
D. 100 watts PEP effective radiated power relative to a dipole

B If an antenna other than a dipole is used, the transmitter output power must be adjusted to account for the gain of the antenna. For example, if the antenna has a gain of 3 dBd (twice the gain of a dipole), the maximum transmitter output power is 25 watts. If an antenna with less gain than a dipole is used, more than 50 watts can be used. See the material on Subelement E9 for more information on effective radiated power (ERP). [97.303(s)]

E1A07 What is the only amateur band where transmission on specific channels rather than a range of frequencies is permitted?

A. 12 meter band
B. 17 meter band
C. 30 meter band
D. 60 meter band

D Operation on 60 meters is restricted to five channels centered on 5332 kHz, 5348 kHz, 5368 kHz, 5373 kHz and 5405 kHz with USB signals having a bandwidth of no more than 2.8 kHz. This secondary allocation to amateurs is restricted in order to maintain compatibility with non-amateur stations who are the primary users of the band. [97.303(s)]

E1A08 What is the only emission type permitted to be transmitted on the 60 meter band by an amateur station?

A. CW
B. RTTY Frequency shift keying
C. Single sideband, upper sideband only
D. Single sideband, lower sideband only

C See the discussion for E1A07. [97.303(s)]

E1A09 Which frequency bands contain at least one segment authorized to only control operators holding an Amateur Extra Class operator license?

A. 80, 75, 40, 20 and 15 meters
B. 80, 40, and 20 meters
C. 75, 40, 30 and 10 meters
D. 160, 80, 40 and 20 meters

A Amateur Extra Class licensees have exclusive operating privileges in the 80, 75, 40, 20 and 15 meter bands. If you look at the frequency allocations, you will see that the 160, 30 and 10 meter bands have no Amateur Extra exclusive segments. [97.301]

E1A10 If a station in a message forwarding system inadvertently forwards a message that is in violation of FCC rules, who is primarily accountable for the rules violation?

A. The control operator of the packet bulletin board station
B. The control operator of the originating station
C. The control operators of all the stations in the system
D. The control operators of all the stations in the system not authenticating the source from which they accept communications

B Amateur message systems are based on trusting that the original messages are proper and legal. Obviously control operators must be responsible for their communications. That's why the rules state, "For stations participating in a message forwarding system, the control operator of the station originating a message is primarily accountable for any violation of the rules in this Part contained in the message."

The rules also state that the control operator of the first forwarding station must either authenticate the identity of the station from which it accepts a communication or accept accountability for any violation of the rules contained in messages it retransmits. [97.219(b), (d)]

E1A11 What is the first action you should take if your digital message forwarding station inadvertently forwards a communication that violates FCC rules?

A. Discontinue forwarding the communication as soon as you become aware of it
B. Notify the originating station that the communication does not comply with FCC rules
C. Notify the nearest FCC Field Engineer's office
D. Discontinue forwarding all messages

A The FCC wants a problem fixed as quickly as possible. The rules say, "Except as noted in paragraph (d) of this section, for stations participating in a message forwarding system, the control operators of forwarding stations that retransmit inadvertently communications that violate the rules in this Part are not accountable for the violative communications. They are, however, responsible for discontinuing such communications once they become aware of their presence." [97.219(c)]

E1A12 If an amateur station is installed on board a ship or aircraft, what condition must be met before the station is operated?

    A. Its operation must be approved by the master of the ship or the pilot in command of the aircraft

    B. The amateur station operator must agree to not transmit when the main ship or aircraft radios are in use

    C. It must have a power supply that is completely independent of the main ship or aircraft power supply

    D. Its operator must have an FCC Marine or Aircraft endorsement on his or her amateur license

A In addition, the amateur station must be separate from all other radio equipment installed on the ship or aircraft, except a common antenna may be shared with a ship installation. Amateur transmissions must not cause interference to other equipment installed on the ship or aircraft and must not be hazardous to the safety of life or property. Finally, there are conditions for stations aboard an aircraft operating under Instrument Flight Rules. [97.11]

E1A13 When a US-registered vessel is in international waters, what type of FCC-issued license or permit is required to transmit amateur communications from an on-board amateur transmitter?

    A. Any amateur license with an FCC Marine or Aircraft endorsement

    B. Any amateur license or reciprocal permit for alien amateur licensee

    C. Only General class or higher amateur licenses

    D. An unrestricted Radiotelephone Operator Permit

B As a licensed US amateur you may operate with your full privileges from a US-registered vessel, but only in international waters. Once in the territorial waters of another country, you are required to abide by their amateur licensing regulations. For the types of reciprocal permits, see Subelement E1F below. You are also required to have permission to operate as in question E1A12. [97.5]

E1B Station restrictions and special operations: restrictions on station location; general operating restrictions, spurious emissions, control operator reimbursement; antenna structure restrictions; RACES operations

E1B01 Which of the following constitutes a spurious emission?

A. An amateur station transmission made at random without the proper call sign identification
B. A signal transmitted in a way that prevents its detection by any station other than the intended recipient
C. Any transmitted bogus signal that interferes with another licensed radio station
D. An emission outside its necessary bandwidth that can be reduced or eliminated without affecting the information transmitted

D Spurious emissions include harmonics, intermodulation or cross-modulation products, or distortion products. These signals are not required components of the desired signal and cause interference to other communications. [97.3(a)(42)]

E1B02 Which of the following factors might cause the physical location of an amateur station apparatus or antenna structure to be restricted?

A. The location is in or near an area of political conflict, military maneuvers or major construction
B. The location's geographical or horticultural importance
C. The location is in an ITU zone designated for coordination with one or more foreign governments
D. The location is significant to our environment, American history, architecture, or culture.

D As you may already know, environmental or historical significance may limit land use in some cases. That's why FCC reflects this concern in Part 97. [97.13(a)]

E1B03 Within what distance must an amateur station protect an FCC monitoring facility from harmful interference?

A. 1 mile
B. 3 miles
C. 10 miles
D. 30 miles

A The Part 97 rule calls for a 1600 meter (1 mile) protection zone. [97.13(b)]

E1B04 What must be done before an amateur station is placed within an officially designated wilderness area or wildlife preserve, or an area listed in the National Register of Historical Places?

A. A proposal must be submitted to the National Park Service
B. A letter of intent must be filed with the National Audubon Society
C. An Environmental Assessment must be submitted to the FCC
D. A form FSD-15 must be submitted to the Department of the Interior

C You only have to deal with FCC. Fortunately, and to simplify matters, you do not have to deal with other governmental or non-governmental agencies. You'll also need to consult Part 1. [97.13(a)]

E1B05 What height restrictions apply to an amateur station antenna structure not close to a public use airport unless the FAA is notified and it is registered with the FCC?

A. It must not extend more than 300 feet above average height of terrain surrounding the site
B. It must be no higher than 200 feet above ground level at its site
C. There are no height restrictions because the structure obviously would not be a hazard to aircraft in flight
D. It must not extend more than 100 feet above sea level or the rim of the nearest valley or canyon

B If the antenna structure is not close to an airport, it can be of any height up to 200 feet above the ground without any FCC or Federal Aviation Administration notification. Once the structure exceeds 200 feet in height, it is subject to FAA rules for lighting, painting and other aviation safety regulations. [97.15(a)]

E1B06 Which of the following additional rules apply if you are installing an amateur station antenna at a site within 20,000 feet of a public use airport?

A. You may have to notify the Federal Aviation Administration and register it with the FCC
B. No special rules apply if your antenna structure will be less than 300 feet in height
C. You must file an Environmental Impact Statement with the EPA before construction begins
D. You must obtain a construction permit from the airport zoning authority

A Don't be fooled because they always seem to refer to Part 97. For this question, you'll also need to consult Part 17. [97.15(a)]

E1B07 Whose approval is required before erecting an amateur station antenna located at or near a public use airport if the antenna would exceed a certain height depending upon the antenna's distance from the nearest active runway?

A. The FAA must be notified and it must be registered with the FCC
B. Approval must be obtained from the airport manager
C. Approval must be obtained from the local zoning authorities
D. The FAA must approve any antenna structure that is higher than 20 feet

A Antennas and antenna towers close to airports can present a safety hazard to aviation. For that reason, these structures must be registered and the FAA notified so that the structure's presence can be communicated to pilots. [97.15(a)]

E1B08 On what frequencies may the operation of an amateur station be restricted if its emissions cause interference to the reception of a domestic broadcast station on a receiver of good engineering design?

A. On the frequency used by the domestic broadcast station
B. On all frequencies below 30 MHz
C. On all frequencies above 30 MHz
D. On the interfering amateur service transmitting frequencies

D Only by avoiding the frequency or frequencies used when the interference occurs would one reduce the interference, and that's reflected in the rules. This assumes that the amateur station is not actually transmitting a spurious signal on the frequency of the broadcast station. [97.121(a)]

E1B09 What is the Radio Amateur Civil Emergency Service (RACES)?

A. A radio service using amateur service frequencies on a regular basis for communications that can reasonably be furnished through other radio services
B. A radio service of amateur stations for civil defense communications during periods of local, regional, or national civil emergencies
C. A radio service using amateur service frequencies for broadcasting to the public during periods of local, regional or national civil emergencies
D. A radio service using local government frequencies by Amateur Radio operators for civil emergency communications

B RACES is, "A radio service using amateur stations for civil defense communications during periods of local, regional or national civil emergencies." [97.3(a)(37)]

E1B10 Which amateur stations may be operated in RACES?

A. Only those club stations licensed to Amateur Extra class operators

B. Any FCC-licensed amateur station except a Technician class operator's station

C. Any FCC-licensed amateur station certified by the responsible civil defense organization for the area served

D. Any FCC-licensed amateur station participating in the Military Affiliate Radio System (MARS)

C Building upon the previous question, RACES operation requires proper registration and an amateur license of any grade. [97.407(a)]

E1B11 What frequencies are normally authorized to an amateur station participating in RACES?

A. All amateur service frequencies otherwise authorized to the control operator

B. Specific segments in the amateur service MF, HF, VHF and UHF bands

C. Specific local government channels

D. Military Affiliate Radio System (MARS) channels

A The key to this question is the phrase "normally authorized" in the question statement. The frequencies that can be used are determined by the control operator license. Under normal circumstances, there are no reserved frequencies for RACES operation. [97.407(b)]

E1B12 What are the frequencies authorized to an amateur station participating in RACES during a period when the President's War Emergency Powers are in force?

A. All frequencies in the amateur service authorized to the control operator

B. Specific amateur service frequency segments authorized in FCC Part 214

C. Specific local government channels

D. Military Affiliate Radio System (MARS) channels

B This question differs from the previous one because of the special circumstances of a national emergency. If there is an emergency that causes the President of the United States to invoke the War Emergency Powers under the Communications Act of 1934, there may be limits placed on the frequencies available for RACES operation. [97.407(b)]

E1B13   What communications are permissible in RACES?

A. Any type of communications when there is no emergency
B. Any Amateur Radio Emergency Service communications
C. Authorized civil defense emergency communications affecting the immediate safety of life and property
D. National defense and security communications authorized by the President

C   Only certain types of communications are permitted in RACES and these must be authorized by the area civil defense organization. Permitted types of communications include national defense and immediate safety of people and property. Authorized training drills and tests are also permitted. [97.407(e)]

E1C   Local, remote and automatic control: Definitions and restrictions pertaining to local, automatic and remote control operation; amateur radio and the Internet; control operator responsibilities for remote and automatically controlled stations

E1C01   What is a remotely controlled station?

A. A station operated away from its regular home location
B. A station controlled by someone other than the licensee
C. A station operating under automatic control
D. A station controlled indirectly through a control link

D   All of these choices may sound like they could fit the definition, but they don't. The key phrase here is operation through a "control link" because the definition is very specific. The rule says that remote control is "The use of a control operator who indirectly manipulates the operating adjustments in the station through a control link to achieve compliance with the FCC Rules." [97.3(a)(38)]

E1C02   What is meant by automatic control of a station?

A. The use of devices and procedures for control so that the control operator does not have to be present at a control point
B. A station operating with its output power controlled automatically
C. Remotely controlling a station's antenna pattern through a directional control link
D. The use of a control link between a control point and a locally controlled station

A   Like remote control, the FCC definition is very precise. The rule says, "The use of devices and procedures for control of a station when it is transmitting so that compliance with the FCC Rules is achieved without the control operator being present at a control point." Note that the compliance with FCC Rules is required. [97.3(a)(6), 97.109(d)]

E1C03 How do the control operator responsibilities of a station under automatic control differ from one under local control?

A. Under local control there is no control operator
B. Under automatic control the control operator is not required to be present at the control point
C. Under automatic control there is no control operator
D. Under local control a control operator is not required to be present at a control point

B   By definition, the control operator of a station operating under automatic control is not required to be physically located at the control point. There must always be a control operator. Local control, as the name implies, requires that the control operator be present to manipulate the transmitter controls directly. [97.3(a)(6), 97.109]

E1C04 When may an automatically controlled station retransmit third party communications?

A. Never
B. Only when transmitting RTTY or data emissions
C. When specifically agreed upon by the sending and receiving stations
D. When approved by the National Telecommunication and Information Administration

B   The rule restricts retransmission of third-party communications to RTTY and data emissions because stations using those modes are more likely to be automated than voice or CW stations. [97.109(e)]

E1C05 When may an automatically controlled station originate third party communications?

A. Never
B. Only when transmitting an RTTY or data emissions
C. When specifically agreed upon by the sending and receiving stations
D. When approved by the National Telecommunication and Information Administration

A   This restriction, stated in the rules as, "All messages that are retransmitted must originate at a station that is being locally or remotely controlled" is intended to prevent commercial and communications-for-hire operation from using Amateur Radio to pass messages. [97.109(e)]

E1C06 Which of the following statements concerning remotely controlled amateur stations is true?

A. Only Extra Class operators may be the control operator of a remote station
B. A control operator need not be present at the control point
C. A control operator must be present at the control point
D. Repeater and auxiliary stations may not be remotely controlled

C   Remote control is assumed to be identical to local control except that the operator is not at the station. The operator is still required to be able to control the transmitter. [97.109(c)]

E1C07 What is meant by local control?

A. Controlling a station through a local auxiliary link
B. Automatically manipulating local station controls
C. Direct manipulation of the transmitter by a control operator
D. Controlling a repeater using a portable handheld transceiver

C   See the discussion for E1C03. [97.3(a)(30)]

E1C08 What is the maximum permissible duration of a remotely controlled station's transmissions if its control link malfunctions?

A. 30 seconds
B. 3 minutes
C. 5 minutes
D. 10 minutes

B   This is why many repeater time-out timers are set to 3 minutes! [97.213(b)]

E1C09 Which of these frequencies are available for automatically controlled ground-station repeater operation?

A. 18.110 - 18.168 MHz
B. 24.940 - 24.990 MHz
C. 10.100 - 10.150 MHz
D. 29.500 - 29.700 MHz

D   Ten meters is the only HF band on which any sort of repeater operation is permitted. [97.205(b)]

E1C10 What types of amateur stations may automatically retransmit the radio signals of other amateur stations?

A. Only beacon, repeater or space stations
B. Only auxiliary, repeater or space stations
C. Only earth stations, repeater stations or model crafts
D. Only auxiliary, beacon or space stations

B Auxiliary stations are often used as remote repeater receivers, providing coverage in difficult terrain. Repeaters and space stations are designed to receive on one frequency, or one set of frequencies, then retransmit any signal they receive. Repeaters do this by demodulating the signal and routing the resulting audio through a separate transmitter. A space station may include a repeater or it may translate signals from one frequency to another without demodulating them. [97.113(f)]

E1D Amateur Satellite service: definitions and purpose; license requirements for space stations; available frequencies and bands; telecommand and telemetry operations; restrictions, and special provisions; notification requirements

E1D01 What is the definition of the term telemetry?

A. One-way transmission of measurements at a distance from the measuring instrument
B. A two-way interactive transmission
C. A two-way single channel transmission of data
D. One-way transmission that initiates, modifies, or terminates the functions of a device at a distance

A This is a definition right out of the FCC rules, which define telemetry as, "A one-way transmission of measurements at a distance from the measuring instrument." [97.3(a)(45)] Practically, "measurements" include any kind of data from the system making the transmissions, such as repeater link status or error codes.

E1D02 What is the amateur-satellite service?

A. A radio navigation service using satellites for the purpose of self-training, intercommunication and technical studies carried out by amateurs
B. A spacecraft launching service for amateur-built satellites
C. A radio communications service using amateur stations on satellites
D. A radio communications service using stations on Earth satellites for weather information gathering

C The amateur satellite service is one of three Amateur Radio services mentioned in Part 97. [97.3(a)(2)] Radio amateurs have built and found launch opportunities for many satellites (called "space stations" in the rules). Transmissions to and from the satellites are carried out on specific sections of the amateur bands.

E1D03 What is a telecommand station in the amateur satellite service?

A. An amateur station located on the Earth's surface for communications with other Earth stations by means of Earth satellites
B. An amateur station that transmits communications to initiate, modify or terminate certain functions of a space station
C. An amateur station located more than 50 km above the Earth's surface
D. An amateur station that transmits telemetry consisting of measurements of upper atmosphere data from space

B You need to know the definition for telecommand station, which is: "An amateur station that transmits communications to initiate, modify, or terminate functions of a space station." [97.3(a)(44)]

E1D04 What is an Earth station in the amateur satellite service?

A. An amateur station within 50 km of the Earth's surface for communications with amateur stations by means of objects in space
B. An amateur station that is not able to communicate using amateur satellites
C. An amateur station that transmits telemetry consisting of measurement of upper atmosphere data from space
D. Any amateur station on the surface of the Earth

A The key here is to notice that the questions deal with stations participating in the satellite service. An Earth station is an amateur station located on, or within 50 km of, the Earth's surface and intended for communications with space stations or with other Earth stations by means of one or more other objects in space. [97.3(a)(16)]

E1D05 What class of licensee is authorized to be the control operator of a space station?

A. Any except those of Technician Class operators
B. Only those of General, Advanced or Amateur Extra Class operators
C. A holder of any class of license
D. Only those of Amateur Extra Class operators

C This may be surprising to you but any valid amateur license holder is eligible to be the control operator of a space station. However that individual is subject to the privileges of the class of operator license held. [97.207(a)]

E1D06 Which of the following special provisions must a space station incorporate in order to comply with space station requirements?

A. The space station must be capable of effecting a cessation of transmissions by telecommand when so ordered by the FCC
B. The space station must cease all transmissions after 5 years
C. The space station must be capable of changing its orbit whenever such a change is ordered by NASA
D. The station call sign must appear on all sides of the spacecraft

A A space station must be capable of ceasing transmissions when ordered to do so by the FCC. As a matter of interest, several amateur satellites have enjoyed operational lives that have exceeded 5 years. [97.207(b)]

E1D07 Which amateur service HF bands have frequencies authorized to space stations?

A. Only 40m, 20m, 17m, 15m, 12m and 10m
B. Only 40m, 20m, 17m, 15m and 10m bands
C. 40m, 30m, 20m, 15m, 12m and 10m bands
D. All HF bands

A Part 97 authorizes space station operation at HF in the 17, 15, 12 and 10-meter bands. It also authorizes operation in parts of the 40 and 20-meter bands. [97.207(c)]

E1D08 Which VHF amateur service bands have frequencies available for space stations?

A. 6 meters and 2 meters
B. 6 meters, 2 meters, and 1.25 meters
C. 2 meters and 1.25 meters
D. 2 meters

D Of the amateur VHF bands, only 2 meters has frequencies authorized for use by space stations. [97.207(c)(2)]

E1D09 Which amateur service UHF bands have frequencies available for a space station?

A. 70 cm
B. 70 cm, 23 cm, 13 cm
C. 70 cm and 33 cm
D. 33 cm and 13 cm

B Now the question moves up to UHF where the 70, 23 and 13-cm bands all have frequencies authorized for space stations. [97.207(c)(2)]

E1D10 Which amateur stations are eligible to be telecommand stations?

A. Any amateur station designated by NASA
B. Any amateur station so designated by the space station licensee
C. Any amateur station so designated by the ITU
D. All of these choices are correct

B The rule says, "Any amateur station designated by the licensee of a space station is eligible to transmit as a telecommand station for that space station, subject to the privileges of the class of operator license held by the control operator." [97.211(a)]

E1D11 Which amateur stations are eligible to operate as Earth stations?

A. Any amateur station whose licensee has filed a pre-space notification with the FCC's International Bureau
B. Only those of General, Advanced or Amateur Extra Class operators
C. Only those of Amateur Extra Class operators
D. Any amateur station, subject to the privileges of the class of operator license held by the control operator

D Your license class is not an impediment to operate an Earth station. Any licensee can do it, within the limitations of the privileges of the class of operator license held by the control operator. [97.209(a)]

E1D12 Who must be notified before launching an amateur space station?

A. The National Aeronautics and Space Administration, Houston, TX
B. The FCC's International Bureau, Washington, DC
C. The Amateur Satellite Corp., Washington, DC
D. All of these answers are correct

B The FCC manages the transmissions of all US owned satellites and so must be notified of satellite launches and status. There are three main notifications: pre-launch, in-space and post-space. The pre-launch deals with preparations for launch and operating plans. Then the FCC wants to know when the satellite goes on and off the air. [97.207(g)]

E1E Volunteer examiner program: definitions, qualifications, preparation and administration of exams; accreditation; question pools; documentation requirements

E1E01 What is the minimum number of qualified VEs required to administer an Element 4 amateur operator license examination?

A. 5
B. 2
C. 4
D. 3

D The rules state, "Each examination for an amateur operator license must be administered by a team of at least 3 VEs at an examination session coordinated by a VEC." There may be more, of course. [97.509(a)]

E1E02 Where are the questions for all written US amateur license examinations listed?

A. In FCC Part 97
B. In an FCC-maintained question pool
C. In the VEC-maintained question pool
D. In the appropriate FCC Report and Order

C The set of questions (one of which you're reading right now) is maintained by all the VECs in a common question pool. Each VEC may choose which questions it uses on exams from the pool (the ARRL VEC uses all of the questions), but all questions on the exams must be from the pool. [97.523]

E1E03 Who is responsible for maintaining the question pools from which all amateur license examination questions must be taken?

A. All of the VECs
B. The VE team
C. The VE question pool team
D. The FCC's Wireless Telecommunications Bureau

A All VECs are required to cooperate in maintaining a single question pool for each written examination element, which they do through their Question Pool Committee (QPC). The common question pool assures an equal treatment of all examinees. The FCC does not maintain the questions. [97.523]

E1E04 What is a Volunteer Examiner Coordinator?

    A. A person who has volunteered to administer amateur operator license examinations

    B. A person who has volunteered to prepare amateur operator license examinations

    C. An organization that has entered into an agreement with the FCC to coordinate amateur operator license examinations

    D. The person that has entered into an agreement with the FCC to be the VE session manager

C A Volunteer Examiner Coordinator is an organization, not a person. The organization enters into an agreement with the FCC to be a VEC. The agreement with the FCC states that the organization exists to further the amateur service, is capable of acting as a VEC, will administer license exams for all classes, and will not discriminate against qualified candidates. [97.521]

E1E05 What is a VE?

    A. An amateur operator who is approved by three or more fellow volunteer examiners to administer amateur license examinations

    B. An amateur operator who is approved by a VEC to administer amateur operator license examinations

    C. An amateur operator who administers amateur license examinations for a fee

    D. An amateur operator who is approved by an FCC staff member to administer amateur operator license examinations

B A VE is simply any amateur who has been certified to be a Volunteer Examiner by one of the VECs. Each VEC is responsible for developing an accreditation system and criteria. [97.525, 97.3(a)(48)]

E1E06 What is a VE team?

    A. A group of at least three VEs who administer examinations for an amateur operator license

    B. The VEC staff

    C. One or two VEs who administer examinations for an amateur operator license

    D. A group of FCC Volunteer Enforcers who investigate Amateur Rules violations

A The rules state that, "Each examination for an amateur operator license must be administered by a team of at least 3 VEs at an examination session coordinated by a VEC." It is that group of at least three Volunteer Examiners that forms the VE team. [97.509]

E1E07 Which of the following persons seeking to become VEs cannot be accredited?

A. Persons holding less than an Advanced Class operator license
B. Persons less than 21 years of age
C. Persons who have ever had an amateur operator or amateur station license suspended or revoked
D. Persons who are employees of the federal government

C The FCC qualifications to be a VE are fairly open. Most amateurs can become a VE without much difficulty. However, the rules specifically exclude any person whose amateur station license or amateur operator license has ever been revoked or suspended. A VE must be at least 18 years of age. [97.509(b)(4)]

E1E08 Which of the following best describes the Volunteer Examiner accreditation process?

A. Each General, Advanced and Amateur Extra Class operator is automatically accredited as a VE when the license is granted
B. The amateur operator applying must pass a VE examination administered by the FCC Enforcement Bureau
C. The prospective VE obtains accreditation from a VE team
D. The procedure by which a VEC confirms that the VE applicant meets FCC requirements to serve as an examiner

D The procedure for accrediting a VE is created by each individual VEC. [97.509(b), 97.525]

E1E09 Where must the VE team be while administering an examination?

A. All of the administering VEs must be present where they can observe the examinees throughout the entire examination
B. The VEs must leave the room after handing out the exam(s) to allow the examinees to concentrate on the exam material
C. The VEs may be elsewhere provided at least one VE is present and is observing the examinees throughout the entire examination
D. The VEs may be anywhere as long as they each certify in writing that examination was administered properly

A Each administering VE must be present and observing the examinees throughout the entire examination. That means every member of the team. The administering VEs are responsible for the proper conduct and necessary supervision of each examination. They can't do that if they're not physically present and actively observing. [97.509(c)]

E1E10 Who is responsible for the proper conduct and necessary supervision during an amateur operator license examination session?

A. The VEC coordinating the session
B. The FCC
C. Each administering VE
D. The VE session manager

C The administering VEs are responsible for the proper conduct and necessary supervision of each examination. It is not just the responsibility of the VE in charge. [97.509(c)]

E1E11 What should a VE do if a candidate fails to comply with the examiner's instructions during an amateur operator license examination?

A. Warn the candidate that continued failure to comply will result in termination of the examination
B. Immediately terminate the candidate's examination
C. Allow the candidate to complete the examination, but invalidate the results
D. Immediately terminate everyone's examination and close the session

B The FCC is very explicit on this point. Discipline will be maintained. The rule says, "The administering VEs must immediately terminate the examination upon failure of the examinee to comply with their instructions." [97.509(c)]

E1E12 To which of the following examinees may a VE not administer an examination?

A. Employees of the VE
B. Friends of the VE
C. The VE's close relatives as listed in the FCC rules
D. All these answers are correct

C Part 97 only restricts VEs from administering examinations to close relatives. It defines these as a spouse, children, grandchildren, stepchildren, parents, grandparents, stepparents, brothers, sisters, stepbrothers, stepsisters, aunts, uncles, nieces, nephews and in-laws. [97.509(d)]

E1E13 What may be the penalty for a VE who fraudulently administers or certifies an examination?

A. Revocation of the VE's amateur station license grant and the suspension of the VE's amateur operator license grant
B. A fine of up to $1000 per occurrence
C. A sentence of up to one year in prison
D. All of these choices are correct

A If you go to Part 97 you will find, "No VE may administer or certify any examination by fraudulent means or for monetary or other consideration including reimbursement in any amount in excess of that permitted. Violation of this provision may result in the revocation of the grant of the VE's amateur station license and the suspension of the grant of the VE's amateur operator license." [97.509(e)]

E1E14 What must the VE team do with the examinee's test papers once they have finished the examination?

A. The VE team must collect and send them to the NCVEC
B. The VE team must collect and send them to the coordinating VEC for grading
C. The VE team must collect and grade them immediately
D. The VE team must collect and send them to the FCC for grading

C Upon completion of each examination element, the VE team is required under the FCC rules to immediately grade the examinee's answers. Further, it is the VE team that is responsible for determining the correctness of the examinee's answers. [97.509(g)]

E1E15 What must the VE team do if an examinee scores a passing grade on all examination elements needed for an upgrade or new license?

A. Photocopy all examination documents and forward them to the FCC for processing
B. Three VEs must certify that the examinee is qualified for the license grant and that they have complied with the VE requirements
C. Issue the examinee the new or upgrade license
D. All these answers are correct

B When the examinee has been credited for all examination elements required for the operator license sought, three VEs must certify that the examinee is qualified for the license grant and that the VEs have complied with the administering VE requirements. [97.509(h)]

E1E16   What must the VE team do with the application form if the examinee does not pass the exam?

A. Return the application document to the examinee
B. Maintain the application form with the VEC's records
C. Send it to the FCC
D. Destroy the application form

A   The VE team must return the application form to the examinee and inform the examinee of the grade. The form may be kept by either the VE team or the examinee. VEC procedures may require the VE team to submit all non-passing application forms with the paperwork for the exam session. [97.509(i)]

E1E17   What are the consequences of failing to appear for re-administration of an examination when so directed by the FCC?

A. The licensee's license will be cancelled
B. The person may be fined or imprisoned
C. The licensee is disqualified from any future examination for an amateur operator license grant
D. All of the above

A   The rule says that the FCC may: Cancel the operator/primary station license of any licensee who fails to appear for re-administration of an examination when directed by the FCC, or who does not successfully complete any required element that is re-administered. In an instance of such cancellation, the person will be granted an operator/primary station license consistent with completed examination elements that have not been invalidated by not appearing for, or by failing, the examination upon re-administration. [97.519(d)(3)]

E1E18   For which types of out-of-pocket expenses may VEs and VECs be reimbursed?

A. Preparing, processing, administering and coordinating an examination for an amateur radio license
B. Teaching an amateur operator license examination preparation course
C. No expenses are authorized for reimbursement
D. Providing amateur operator license examination preparation training materials

A   VEs cannot charge for exams. However, the fee charged may cover out-of-pocket and other administrative fees — a subtle difference. The rule says, "VEs and VECs may be reimbursed by examinees for out-of-pocket expenses incurred in preparing, processing, administering, or coordinating an examination for an amateur operator license." [97.527]

E1E19 How much reimbursement may the VE team and VEC accept for preparing, processing, administering and coordinating an examination?

A. Actual out-of-pocket expenses
B. The national minimum hourly wage for time spent providing examination services
C. Up to the maximum fee per examinee announced by the FCC annually
D. As much as the examinee is willing to donate

A   If you become a VE, do not expect to make money from the experience, but you can recoup your expenses. You can't charge the maximum fee unless you have expenses in that amount or more. You may not keep any fees other than those to pay for actual out-of-pocket expenses. [97.509(e), 97.527]

E1E20 What is the minimum age to be a volunteer examiner?

A. 13 years old
B. 16 years old
C. 18 years old
D. 21 years old

C   There are very few restrictions as to who can become a VE, but a minimum age is required to be sure that the VEs are capable of carrying out their duties. This is important when applicants of all ages may be present to take an exam. [97.509(b)]

E1F   Miscellaneous rules: external RF power amplifiers; Line A; national quiet zone; business communications; compensated communications; spread spectrum; auxiliary stations; reciprocal operating privileges; IARP and CEPT licenses; third party communications with foreign countries; special temporary authority

E1F01 On what frequencies are spread spectrum transmissions permitted?

A. Only on amateur frequencies above 50 MHz
B. Only on amateur frequencies above 222 MHz
C. Only on amateur frequencies above 420 MHz
D. Only on amateur frequencies above 144 MHz

B   Because of the necessary bandwidth of spread spectrum signals, they are permitted only on the 222 MHz and higher frequency bands. [97.305]

E1F02 Which of the following operating arrangements allows an FCC-licensed US citizen to operate in many European countries, and alien amateurs from many European countries to operate in the US?

A. CEPT agreement
B. IARP agreement
C. ITU reciprocal license
D. All of these choices are correct

A CEPT is the European Conference of Posts and Telecommunications. The US is a participant in the CEPT Recommendation, which allows US amateurs to operate in certain European countries. It also allows amateur from many European countries to operate in the US. [97.5(d)]

E1F03 Which of the following operating arrangements allow an FCC-licensed US citizen and many Central and South American amateur operators to operate in each other's countries?

A. CEPT agreement
B. IARP agreement
C. ITU agreement
D. All of these choices are correct

B The IARP is the International Amateur Radio Permit, which covers countries in the Americas. The agreement came out of CITEL (Inter-American Telecommunication Commission). [97.5(e)]

E1F04 What does it mean if an external RF amplifier is listed on the FCC database as certificated for use in the amateur service?

A. The RF amplifier may be marketed for use in any radio service
B. That particular RF amplifier may be marketed for use in the amateur service
C. All similar RF amplifiers produced by other manufacturers may be marketed
D. All RF amplifiers produced by that manufacturer may be marketed

B Certification is for a specific model and manufacturer. Certification means that it may be marketed for use in the amateur service. [97.315(c)]

E1F05   Under what circumstances may a dealer sell an external RF power amplifier capable of operation below 144 MHz if it has not been granted FCC certification?

A. It was purchased in used condition from an amateur operator and is sold to another amateur operator for use at that operator's station
B. The equipment dealer assembled it from a kit
C. It was imported from a manufacturer in a country that does not require certification of RF power amplifiers
D. It was imported from a manufacturer in another country, and it was certificated by that country's government

A   There are three conditions that must be satisfied. First, the amplifier must be in "used condition." Second, it must be purchased from an amateur operator and sold to another amateur operator. (The wording allows an equipment dealer to be involved.) Third, the amplifier must be for use at the purchasing operator's station. [97.315(b)(3)]

E1F06   Which of the following geographic descriptions approximately describes "Line A"?

A. A line roughly parallel to and south of the US-Canadian border
B. A line roughly parallel to and west of the US Atlantic coastline
C. A line roughly parallel to and north of the US-Mexican border and Gulf coastline
D. A line roughly parallel to and east of the US Pacific coastline

A   If you operate on the 70-cm band in the northern states, you need to know about Line A which parallels the US-Canadian border. [97.3(a)(29)] Line A exists to protect Canadian UHF non-amateurs using the 420-430 MHz band from interference by US amateur signals.

E1F07   Amateur stations may not transmit in which of the following frequency segments if they are located north of Line A?

A. 440 - 450 MHz.
B. 53 - 54 MHz
C. 222 - 223 MHz
D. 420 - 430 MHz

D   Line A is important because there is a restriction at UHF. The rules say, "No amateur station shall transmit from north of Line A in the 420-430 MHz segment." [97.303(f)(1)]

E1F08 **What is the National Radio Quiet Zone?**

A. An area in Puerto Rico surrounding the Aricebo Radio Telescope
B. An area in New Mexico surrounding the White Sands Test Area
C. An area surrounding the National Radio Astronomy Observatory
D. An area in Florida surrounding Cape Canaveral

C The National Radio Quiet Zone protects the radio astronomy facilities for the National Radio Astronomy Observatory in Green Bank, WV and the Naval Research Laboratory at Sugar Grove, WV. It is composed of parts of the states of Maryland, West Virginia and Virginia. [97.3(a)(32)]

E1F09 **When may the control operator of a repeater accept payment for providing communication services to another party?**

A. When the repeater is operating under portable power
B. When the repeater is operating under local control
C. During Red Cross or other emergency service drills
D. Under no circumstances

D This falls into the area of prohibited transmissions. Specifically prohibited are: "Communications for hire or for material compensation, direct or indirect, paid or promised, except as otherwise provided in these rules." There is no exception in the rules for repeater control operator services. [97.113(a)(2)]

E1F10 **When may an amateur station send a message to a business?**

A. When the total money involved does not exceed $25
B. When the control operator is employed by the FCC or another government agency
C. When transmitting international third-party communications
D. When neither the amateur nor his or her employer has a pecuniary interest in the communications

D This also falls into the area of prohibited transmissions. The rule prohibits, "Communications in which the station licensee or control operator has a pecuniary interest, including communications on behalf of an employer." [97.113(a)(3)]

E1F11 Which of the following types of amateur-operator-to-amateur-operator communications are prohibited?

A. Communications transmitted for hire or material compensation, except as otherwise provided in the rules
B. Communications that have a political content, except as allowed by the Fairness Doctrine
C. Communications that have a religious content
D. Communications in a language other than English

A   See the discussion for E1F09. [97.113]

E1F12 FCC-licensed amateur stations may use spread spectrum (SS) emissions to communicate under which of the following conditions?

A. When the other station is in an area regulated by the FCC
B. When the other station is in a country permitting SS communications
C. When the transmission is not used to obscure the meaning of any communication
D. All of these choices are correct

D   All of the statements are correct. Here's the rule: "SS emission transmissions by an amateur station are authorized only for communications between points within areas where the amateur service is regulated by the FCC and between an area where the amateur service is regulated by the FCC and an amateur station in another country that permits such communications. SS emission transmissions must not be used for the purpose of obscuring the meaning of any communication." [97.311(a)]

E1F13 What is the maximum transmitter power for an amateur station transmitting spread spectrum communications?

A. 1 W
B. 1.5 W
C. 100 W
D. 1.5 kW

C   The SS transmitter power must not exceed 100 W under any circumstances. [97.311(d)] This is to keep the "noise" from SS communications from causing interference to users of single-channel modes.

E1F14 Which of the following best describes one of the standards that must be met by an external RF power amplifier if it is to qualify for a grant of FCC certification?

A. It must produce full legal output when driven by not more than 5 watts of mean RF input power
B. It must be capable of external RF switching between its input and output networks
C. It must exhibit a gain of 0 dB or less over its full output range
D. It must satisfy the FCC's spurious emission standards when operated at its full output power

D The FCC rules state "To receive a grant of certification, the amplifier must satisfy the spurious emission standards...when the amplifier is operated at the lesser of 1.5 kW PEP or its full output power and when the amplifier is placed in the 'standby' or 'off' positions while connected to the transmitter." [97.317(a)(1)]

E1F15 Who may be the control operator of an auxiliary station?

A. Any licensed amateur operator
B. Only Technician, General, Advanced or Amateur Extra Class operators
C. Only General, Advanced or Amateur Extra Class operators
D. Only Amateur Extra Class operators

B Only Novice class licensees are not authorized to be the control operator of an auxiliary station. [97.201(a)]

E1F16 What types of communications may be transmitted to amateur stations in foreign countries?

A. Business-related messages
B. Automatic retransmissions of any amateur communications
C. Communications incidental to the purpose of the amateur service and remarks of a personal nature
D. All of these choices are correct

C Even third-party communications are subject to the restrictions of non-commercial content. Although today the Internet provides an outlet for the disallowed types of communication, these rules were put in place at a time when few alternative services were available. It is still important to keep Amateur Radio free of commercial content. [97.117]

E1F17 Under what circumstances might the FCC issue a "Special Temporary Authority" (STA) to an amateur station?

A. To provide for experimental amateur communications
B. To allow regular operation on Land Mobile channels
C. To provide additional spectrum for personal use
D. To provide temporary operation while awaiting normal licensing

A While uncommon, STAs are granted in order for amateurs to experiment with unusual or new modes. For example, STAs were issued to allow amateurs to experiment with spread-spectrum techniques before a formal set of rules was put in place for all amateurs. [1.931]

Operating Practices and Procedures

There will be five questions on your Extra class examination from the Operating Procedures subelement. These five questions will be taken from the five groups of questions labeled E2A through E2E.

E2A Amateur radio in space: amateur satellites; orbital mechanics; frequencies and modes; satellite hardware; satellite operations

E2A01 What is the direction of an ascending pass for an amateur satellite?

A. From west to east
B. From east to west
C. From south to north
D. From north to south

C If the satellite is moving from south to north as it passes over your area, then it is making an ascending pass.

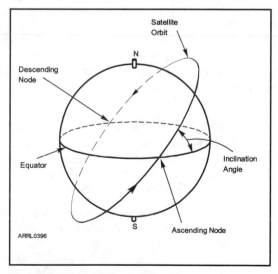

Figure E2A01 — This drawing illustrates basic satellite orbital terminology.

E2A02   What is the direction of a descending pass for an amateur satellite?

A. From north to south
B. From west to east
C. From east to west
D. From south to north

A   If the satellite is moving from north to south as it passes over your area, then it is making a descending pass.

E2A03   What is the orbital period of a satellite?

A. The point of maximum height of a satellite's orbit
B. The point of minimum height of a satellite's orbit
C. The time it takes for a satellite to complete one revolution around the Earth
D. The time it takes for a satellite to travel from perigee to apogee

C   The period of a satellite is the time it takes that satellite to complete one orbit.

E2A04   What is meant by the term "mode" as applied to an amateur radio satellite?

A. The type of signals that can be relayed through the satellite
B. The satellite's uplink and downlink frequency bands
C. The satellite's orientation with respect to the Earth
D. Whether the satellite is in a polar or equatorial orbit

B   The word "mode" is a bit overloaded when it comes to satellites. When referring to the satellite itself and not the type of signals it relays, mode refers to how the satellite's uplink and downlink frequencies are configured. Some satellites have only one mode, while others with multiple receivers and transmitters may be able to operate in several modes.

E2A05   What do the letters in a satellite's mode designator specify?

A. Power limits for uplink and downlink transmissions
B. The location of the ground control station
C. The polarization of uplink and downlink signals
D. The uplink and downlink frequencies

D   Each frequency band is abbreviated by a single letter: VHF by V, UHF by U and so forth. The first letter specifies the uplink and the second letter specifies the downlink. See Table E2-1.

Table E2-1
Satellite Operating Modes

Mode	Satellite Receive (Uplink)	Satellite Transmit (Downlink)
V/H	VHF (144 - 146 MHz)	HF (21 - 30 MHz)
U/V	UHF (435 - 438 MHz)	VHF (144 - 146 MHz)
V/U	VHF (144 - 146 MHz)	UHF (435 - 438 MHz)
L/U	L-Band (1.26 - 1.27 GHz)	UHF (435 - 438 MHz)

E2A06 On what band would a satellite receive signals if it were operating in mode U/V?

- A. 432 MHz
- B. 144 MHz
- C. 50 MHz
- D. 28 MHz

A The first letter of a mode descriptor indicates the uplink frequency band, so U/V is the UHF uplink and VHF downlink mode. Although the 432 MHz band is the correct answer, the actual frequencies used for satellite uplinks are in the 435-438 MHz range. See Table E2-1.

E2A07 Which of the following types of signals can be relayed through a linear transponder?

- A. FM and CW
- B. SSB and SSTV
- C. PSK and Packet
- D. All these answers are correct

D By convention, transponder is the name given to a linear translator that is installed in a satellite. It is somewhat like a repeater in that both devices receive signals and retransmit them. A repeater does that for signals of a single mode on a single frequency. By contrast, a transponder's receive passband includes enough spectrum for many channels. The satellite transponder translates (or converts the frequency of) all signals in its passband — regardless of mode. The transponder then amplifies them and retransmits them in the new frequency range.

E2A08 What is the primary reason for satellite users to limit their transmit ERP?

A. For RF exposure safety
B. Because the satellite transmitter output power is limited
C. To avoid limiting the signal of the other users
D. To avoid interfering with terrestrial QSOs

B Satellites have a very limited power supply consisting of solar cells and batteries. This means that the satellite downlink (transmit) transmitter power must be limited. Most satellites use linear transponders (see E2A07) in which a strong signal on the uplink is also a strong signal on the downlink. If a satellite user's ERP is too high, it can consume a disproportionate amount of the available transmit power. For that reason, satellite users should use the minimum transmitter output power needed to communicate through the satellite.

E2A09 What do the terms L band and S band specify with regard to satellite communications?

A. The 23 centimeter and 13 centimeter bands
B. The 2 meter and 70 centimeter bands
C. FM and Digital Store-and-Forward systems
D. Which sideband to use

A Instead of wavelength, microwave bands are designated by a letter. The Amateur Radio 23-cm band (1296 MHz) is part of the industry-designated L-band. The Amateur Radio 13-cm band (2.4 GHz) is part of the industry-designated S-band.

E2A10 Why may the received signal from an amateur satellite exhibit a rapidly repeating fading effect?

A. Because the satellite is rotating
B. Because of ionospheric absorption
C. Because of the satellite's low orbital altitude
D. Because of the Doppler effect

A Satellite designers usually spin a satellite to improve the stability of its orientation. Of course the satellite antennas spin too, and this results in a fairly rapid pulsed fading effect. This effect is called spin modulation.

E2A11 What type of antenna can be used to minimize the effects of spin modulation and Faraday rotation?

A. A linearly polarized antenna
B. A circularly polarized antenna
C. An isotropic antenna
D. A log-periodic dipole array

B Circularly polarized antennas of the proper sense (direction of polarization rotation) will minimize the effects of spin modulation described in the discussion for E2A10. The polarization of a radio signal passing through the ionosphere does not remain constant. A horizontally polarized signal leaving a satellite will not be horizontally polarized after it passes through the ionosphere on its way to Earth. That signal will in fact seem to be changing polarization at a receiving station. This effect is called Faraday rotation. The best way to compensate for Faraday rotation is to use circularly polarized antennas for transmitting and receiving.

E2A12 What is one way to predict the location of a satellite at a given time?

A. By means of the Doppler data for the specified satellite
B. By subtracting the mean anomaly from the orbital inclination
C. By adding the mean anomaly to the orbital inclination
D. By calculations using the Keplerian elements for the specified satellite

D Johannes Kepler described the planetary orbits of our solar system. The laws and mathematical formulas that he developed may be used to calculate the location of a satellite at a given time. By using as input the values of a set of measurements describing the satellite orbit, called Keplerian elements, computer software can determine the location of the satellite.

E2A13 What type of satellite appears to stay in one position in the sky?

A. HEO
B. Geosynchronous
C. Geomagnetic
D. LEO

B The orbital period of a geosynchronous satellite is the same as the Earth's period of rotation about its axis. As a result, the satellite stays above the same spot on the Earth's surface, revolving around the Earth's axis at the same rate as the Earth's rotation.

E2A14 What happens to a satellite's transmitted signal due to the Doppler Effect?

A. The signal strength is reduced as the satellite passes overhead
B. The signal frequency shifts lower as the satellite passes overhead
C. The signal frequency shifts higher as the satellite passes overhead
D. The polarization of the signal continually rotates

B The Doppler Effect (or Doppler shift) describes the way the downlink frequency of a satellite varies by several kHz during a low-earth orbit. Doppler shift is caused by the relative motion between you and the satellite. As the satellite is moving toward you, the frequency of its downlink signal appears to increase by a small amount. When the satellite passes overhead and begins to move away from you, the downlink signal drops in much the same way as the tone of a car horn or train whistle drops as the vehicle moves past you.

E2B Television practices: fast scan television standards and techniques; slow scan television standards and techniques

E2B01 How many times per second is a new frame transmitted in a fast-scan (NTSC) television system?

A. 30
B. 60
C. 90
D. 120

A A picture is divided sequentially into pieces for transmission or viewing; this process is called scanning. A total of 525 scan lines comprise a frame (complete picture) in the analog/NTSC US television system. Thirty frames are generated each second.

E2B02 How many horizontal lines make up a fast-scan (NTSC) television frame?

A. 30
B. 60
C. 525
D. 1080

C A total of 525 horizontal lines make up a fast-scan television frame (complete picture).

E2B03 How is an interlace scanning pattern generated in a fast-scan (NTSC) television system?

A. By scanning two fields simultaneously
B. By scanning each field from bottom to top
C. By scanning lines from left to right in one field and right to left in the next
D. By scanning odd numbered lines in one field and even numbered ones in the next

D The fast-scan TV frame consists of two fields of 262½ lines each. The odd numbered lines are scanned in one field and the even numbered ones are scanned in the next. The half line is the secret to the interlaced scan pattern.

E2B04 What is blanking in a video signal?

A. Synchronization of the horizontal and vertical sync pulses
B. Turning off the scanning beam while it is traveling from right to left or from bottom to top
C. Turning off the scanning beam at the conclusion of a transmission
D. Transmitting a black and white test pattern

B The function of blanking is to turn off the scanning beam while the beam is traveling from right to left and from bottom to top. The blanking signal occurs at the end of each scan line and at the end of each scan field. In terms of video brightness, blanking is "blacker than black" so that the blanked beam will not affect the image as it travels across it.

E2B05 Which of the following is an advantage of using vestigial sideband for standard fast scan TV transmissions?

A. The vestigial sideband carries the audio information
B. The vestigial sideband contains chroma information
C. Vestigial sideband reduces bandwidth while allowing for simple video detector circuitry
D. Vestigial sideband provides high frequency emphasis to sharpen the picture

C Including a vestigial sideband (see the next question) allows the video signal to be recovered with simple AM detector circuits, as opposed to requiring the more complex SSB demodulators. By only including part of the unused sideband, the full bandwidth of an AM-DSB signal is not required.

E2B06   What is vestigial sideband modulation?

A. Amplitude modulation in which one complete sideband and a portion of the other sideband is transmitted
B. A type of modulation in which one sideband is inverted
C. Narrow-band FM transmission achieved by filtering one sideband from the audio before frequency modulating the carrier
D. Spread spectrum modulation achieved by applying FM modulation following single sideband amplitude modulation

A   Vestigial means partial and unused, and the vestigial sideband of a fast-scan TV signal is attenuated and reduced in bandwidth from the full-bandwidth sideband that carries the video information. Its function is to allow the transmitted signal to be detected as an AM signal and limit transmission bandwidth.

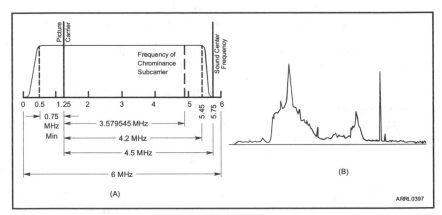

Figure E2B06 — The frequency spectrum of a color fast-scan TV signal is shown in part A. The vestigial sideband is to the left of the line labeled "Picture Carrier" and occupies a bandwidth of 1.25 MHz. Part B shows the spectrum analyzer view of a typical fast-scan TV signal.

E2B07   What is the name of the video signal component that carries color information?

A. Luminance
B. Chroma
C. Hue
D. Spectral Intensity

B   Also referred to as chrominance, the chroma signal is combined with the basic monochrome (black and white) TV signal.

E2B08 Which of the following is a common method of transmitting accompanying audio with amateur fast-scan television?

A. Frequency-modulated sub-carrier
B. A separate VHF or UHF audio link
C. Frequency modulation of the video carrier
D. All of these choices are correct

D The first choice is how commercial stations transmit the audio information in a TV signal. The second is the way most hams send ATV audio because it is easier than adding an FM signal to the transmitted video signal. The third choice is another popular method of adding audio information without affecting the AM video components.

E2B09 What hardware, other than a transceiver with SSB capability and a suitable computer, is needed to decode SSTV based on Digital Radio Mondiale (DRM)?

A. A special IF converter
B. A special front end limiter
C. A special notch filter to remove synchronization pulses
D. No other hardware is needed

D Digital Radio Mondiale was developed for transmitting digital audio information and data over shortwave broadcast channels. Hams adapted it to transmit SSTV pictures as digital data. The demodulated audio from a DRM signal can be processed by computer sound card, recovering the picture information.

E2B10 Which of the following is an acceptable bandwidth for Digital Radio Mondiale (DRM) based voice or SSTV digital transmissions made on the HF amateur bands?

A. 3 kHz
B. 10 kHz
C. 15 kHz
D. 20 kHz

A Image transmissions are allowed in phone band segments if their bandwidth is no greater than that of a voice signal of the same modulation type. For SSB transmissions, the normal bandwidth is 3 kHz.

E2B11 What is the function of the Vertical Interval Signaling (VIS) code transmitted as part of an SSTV transmission?

A. To lock the color burst oscillator in color SSTV images
B. To identify the SSTV mode being used
C. To provide vertical synchronization
D. To identify the callsign of the station transmitting

B There are a number of different formats in which SSTV images are transmitted. So that automated receiving systems can determine which format is being used, identifying codes are transmitted during the vertical blanking period between images. This is called Vertical Interval Signaling (VIS).

E2B12 How are analog slow-scan television images typically transmitted on the HF bands?

A. Video is converted to equivalent Baudot representation
B. Video is converted to equivalent ASCII representation
C. Varying tone frequencies representing the video are transmitted using FM
D. Varying tone frequencies representing the video are transmitted using single sideband

D The standard method of transmitting analog SSTV on the HF bands uses an SSB transmitter. The raster control signals, such as horizontal and vertical sync pulses, and image video are encoded as different tones that are transmitted as regular audio over the SSB channel.

E2B13 How many lines are commonly used in each frame on an amateur slow-scan color television picture?

A. 30 to 60
B. 60 or 100
C. 128 or 256
D. 180 or 360

C Color SSTV standards specify either 128 or 256 lines per frame depending on the format that you chose.

E2B14 What aspect of an amateur slow-scan television signal encodes the brightness of the picture?

A. Tone frequency
B. Tone amplitude
C. Sync amplitude
D. Sync frequency

A Because HF signals experience frequent fading, amplitude is not used to encode the SSTV signal. Frequency of the received tones is unaffected by fading, however, and can be used to encode brightness or any control function of the signal.

E2B15 What signals SSTV receiving equipment to begin a new picture line?

A. Specific tone frequencies
B. Elapsed time
C. Specific tone amplitudes
D. A two-tone signal

A See the discussion for E2B12.

E2B16 Which of the following is the video standard used by North American Fast Scan ATV stations?

A. PAL
B. DRM
C. Scottie
D. NTSC

D NTSC stands for National Television Standard Committee and is the standard for analog commercial television broadcasters, as well. The NTSC standard describes all aspects of the analog television signal.

E2B17 Which of the following is NOT a characteristic of FMTV (Frequency-Modulated Amateur Television) as compared to vestigial sideband AM television?

A. Immunity from fading due to limiting
B. Poor weak signal performance
C. Greater signal bandwidth
D. Greater complexity of receiving equipment

A You are looking for the false statement. FMTV, like other FM transmission systems, has no immunity from fading due to limiting in the FM receiver. The other statements are true characteristics of FMTV as compared to vestigial sideband AM television.

E2B18   What is the approximate bandwidth of a slow-scan TV signal?

A.   600 Hz
B.   3 kHz
C.   2 MHz
D.   6 MHz

B   See the discussion for E2B10.

E2B19   On which of the following frequencies is one likely to find FMTV transmissions?

A.   14.230 MHz
B.   29.6 MHz
C.   52.525 MHz
D.   1255 MHz

D   Because the bandwidth of FMTV signals is so wide (approximately 12 MHz), its use is restricted to UHF and higher bands.

E2B20   What special operating frequency restrictions are imposed on slow scan TV transmissions?

A.   None; they are allowed on all amateur frequencies
B.   They are restricted to 7.245 MHz, 14.245 MHz, 21.345, MHz, and 28.945 MHz
C.   They are restricted to phone band segments and their bandwidth can be no greater than that of a voice signal of the same modulation type
D.   They are not permitted above 54 MHz

C   See the discussion for E2B10.

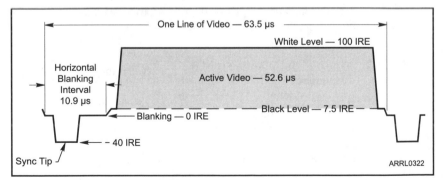

Figure E2B21 — The transmitted video waveform is measured in IRE units because the absolute voltage level varies with transmitter power.

Table E2-2

Standard Video Levels

	IRE Units
Zero carrier	120
White	100
Black	7.5
Blanking	0
Sync tip	−40

E2B21 If 100 IRE units correspond to the most-white level in the NTSC standard video format, what is the level of the most-black signal?

A. 140 IRE units
B. 7.5 IRE units
C. 0 IRE units
D. −40 IRE units

B IRE (or IEEE units) are used to measure levels in a standard video signal. (See Table E2-2). Percent of PEV (Peak Envelope Voltage) is also used, but is an obsolete method. The figure shows the standard IRE units for the levels of several aspects of the signal.

E2C Operating methods, part 1: contest and DX operating; spread-spectrum transmissions; automatic HF forwarding; selecting an operating frequency

E2C01 Which of the following is true about contest operating?

A. Operators are permitted to make contacts even if they do not submit a log
B. Interference to other amateurs is unavoidable and therefore acceptable
C. It is mandatory to transmit the call sign of the station being worked as part of every transmission to that station
D. Every contest requires a signal report in the exchange

A With very few exceptions, contests are open to any amateur who happens to tune across them. In addition, stations participating in the contest may count contacts with you even if you do not send the event sponsor a log of your contest activity. Just ask the contest station what information they need to complete the contact and they'll tell you.

E2C02 Which of the following best describes "self spotting" in regards to contest operating?

A. The generally prohibited practice of posting one's own call sign and frequency on a call sign spotting network
B. The acceptable practice of manually posting the call signs of stations on a call sign spotting network
C. A manual technique for rapidly zero beating or tuning to a station's frequency before calling that station
D. An automatic method for rapidly zero beating or tuning to a station's frequency before calling that station

A A "spot" is an announcement that a specific station is present on a specific frequency. Spots are distributed mostly over the Internet via Web sites and TELNET-based "clusters" that form a worldwide spotting network. Posting your own frequency and call sign — self-spotting — in an attempt to get more contacts is prohibited.

E2C03 From which of the following bands is amateur radio contesting generally excluded?

A. 30 meters
B. 6 meters
C. 2 meters
D. 33 cm

A By general agreement, 60, 30, 17 and 12 meters are excluded from contest activity. This gives stations not interested in the contest an additional option for avoiding contest activity.

E2C04 On which of the following frequencies is an amateur radio contest contact generally discouraged?

A. 3.525 MHz
B. 14.020 MHz
C. 28.330 MHz
D. 146.52 MHz

D The national simplex calling channels of 146.52, 223.5 and 446.0 MHz are generally excluded from contest activity. This keeps them useful for establishing local contacts.

E2C05 Which of the following frequencies would generally be acceptable for U.S. stations to work other U.S. stations in a phone contest?

A. 5405 kHz
B. 14.310 MHz
C. 50.050 MHz
D. 146.52 MHz

B 14.310 MHz is a general-use frequency without other nationally agreed on uses. Frequencies A and D are excluded from contest activity by convention. Phone activity of any sort is not permitted on frequency C. [97.301]

E2C06 During a VHF/UHF contest, in which band segment would you expect to find the highest level of activity?

A. At the top of each band, usually in a segment reserved for contests
B. In the middle of each band, usually on the national calling frequency
C. In the weak signal segment of the band, with most of the activity near the calling frequency
D. In the middle of the band, usually 25 kHz above the national calling frequency

C Most contest activity on the VHF and UHF bands is SSB or CW, referred to as "weak signal" because these modes perform better than FM at low signal-to-noise ratios. The usual convention for contest activity is to call CQ on or near the calling frequency and then move to a nearby frequency once activity builds. In populated areas, calling CQ on the calling frequency is discouraged so that stations will spread out.

E2C07 What is the Cabrillo format?

A. A standard for organizing information in contest log files
B. A method of exchanging information during a contest QSO
C. The most common set of contest rules
D. The rules of order for meetings between contest sponsors

A Named for Cabrillo College near Santa Cruz, California, the Cabrillo format specifies how log information in text form is organized for submission to the contest sponsor. It does not control how logs are checked or scored by the sponsor.

E2C08   Why are received spread-spectrum signals resistant to interference?

A. Signals not using the spectrum-spreading algorithm are suppressed in the receiver
B. The high power used by a spread-spectrum transmitter keeps its signal from being easily overpowered
C. The receiver is always equipped with a digital blanker circuit
D. If interference is detected by the receiver it will signal the transmitter to change frequencies

A   Because the spread-spectrum signal changes frequencies rapidly, interference or noise on a single frequency affects it only briefly. To interfere with a spread-spectrum signal, interference would have to follow the changing frequency exactly or cover a significant fraction of the band it occupies.

E2C09   How does the spread-spectrum technique of frequency hopping (FH) work?

A. If interference is detected by the receiver it will signal the transmitter to change frequencies
B. If interference is detected by the receiver it will signal the transmitter to wait until the frequency is clear
C. A pseudo-random binary bit stream is used to shift the phase of an RF carrier very rapidly in a particular sequence
D. The frequency of the transmitted signal is changed very rapidly according to a particular sequence also used by the receiving station

D   FH spread spectrum is a form of spreading in which the center frequency of a conventional carrier is altered many times per second in accordance with a list of frequency channels. The same frequency list is used by the receiving station, so it "hops" in sync with the transmitter.

E2C10   Why might a phone DX station state that he is listening on another frequency?

A. Because the DX station may be transmitting on a frequency that is prohibited to some responding stations
B. To separate the calling stations from the DX station
C. To reduce interference, thereby improving operating efficiency
D. All of these choices are correct

D   By transmitting on one frequency and listening on another (split frequency operation) when many stations are calling, it is much easier for the callers to hear the DX station. This allows everyone to follow instructions and makes the operation much more efficient. If you hear a DX station working callers you can't hear, tune around to see if you can find the callers on a nearby frequency — usually above that of the DX station.

E2C11 How should you generally sign your call when attempting to contact a DX station working a "pileup" or in a contest?

A. Send your full call sign once or twice
B. Send only the last two letters of your call sign until you make contact
C. Send your full call sign and grid square
D. Send the call sign of the DX station three times, the words "this is", then your call sign three times

A When the other operator sends CQ or QRZ? (Who is calling me?), this is your cue to call by sending your full call sign. Sending only part of your call sign is not a proper way to identify your station.

E2C12 In North America during low sunspot activity, when signals from Europe become weak and fluttery across an entire HF band two to three hours after sunset, what might help to contact other European DX stations?

A. Switch to a higher frequency HF band
B. Switch to a lower frequency HF band
C. Wait 90 minutes or so for the signal degradation to pass
D. Wait 24 hours before attempting another communication on the band

B The "weak and fluttery after sunset" condition indicates that the band conditions are deteriorating. You can continue to operate by changing frequencies. In this case, you'll want to operate at a lower frequency because the MUF has moved lower as well.

E2D   Operating methods, part 2: VHF and UHF digital modes; packet clusters; Automatic Position Reporting System (APRS)

E2D01 What does "command mode" mean in packet operations?

A. Your computer is ready to run packet communications software
B. The TNC is ready to receive instructions via the keyboard
C. Your TNC has received a command packet from a remote TNC
D. The computer is ready to be set up to communicate with the TNC

B Command mode means that the TNC (terminal node controller) is ready to accept commands from an operator to change modes, initiate or terminate a connection, or perform another function.

E2D02 What is the definition of "baud"?

A. The number of data symbols transmitted per second
B. The number of characters transmitted per second
C. The number of characters transmitted per minute
D. The number of words transmitted per minute

A A baud is a single signaling event between the data link's transmitter and receiver. Each signaling event transmits one symbol. A symbol can encode one or more bits. The rate at which signaling events occur is called baud or bauds.

E2D03 Which of the follow is true when comparing HF and 2-meter packet operations?

A. HF packet typically uses FSK with a data rate of 300 baud; 2-meter packet uses AFSK with a data rate of 1200 baud
B. HF packet and 2-meter packet operations use different codes for information exchange
C. HF packet is limited to Amateur Extra class amateur licensees; 2-meter packet is open to all but Novice Class amateur licensees
D. HF and 2-meter packet operations are both limited to CW/Data-only band segments

A HF and 2-meter packet modes both use the same AX.25 protocol to control the exchange of data, but they use different data rates and modulation. HF packet is transmitted using FSK SSB and is limited to 300 bauds by regulation. It rarely achieves that full rate because of the nature of HF propagation. On 2 meters, packet uses AFSK FM and most commonly operates at 1200 bauds.

E2D04 What is the purpose of digital store-and-forward functions on an Amateur satellite?

A. To upload operational software for the transponder
B. To delay download of telemetry until the satellite is over the control station
C. To store digital messages in the satellite for later download by other stations
D. To relay messages between satellites

C Digital communications satellites provide non-real-time computer-to-computer communications. These satellites work like temporary mailboxes in space. You upload a message or a file to a satellite and it is stored for a time (could be days or weeks) until someone else — possibly on the other side of the world — downloads it. Messages or files can be sent to an individual recipient or everyone. Such communications are called store-and-forward.

E2D05 Which of the following techniques is normally used by low-earth orbiting digital satellites to relay messages around the world?

A. Digipeating
B. Store-and-forward
C. Multi-satellite relaying
D. Node hopping

B Since low-earth orbiting (LEO) satellites cannot see large segments of the world at once, messages are sent via store-and-forward mode which allows recipients to pick up their messages from the satellite mailbox as the satellite flies overhead. See also E2D04.

E2D06 Which of the following is a commonly used 2-meter APRS frequency?

A. 144.20 MHz
B. 144.39 MHz
C. 145.02 MHz
D. 146.52 MHz

B The common 2-meter Automatic Position Reporting System (APRS) frequency in North America is 144.39 MHz.

E2D07 Which of the following digital protocols is used by APRS?

A. AX.25
B. 802.11
C. PACTOR
D. AMTOR

A The APRS system is constructed to use the amateur AX.25 packet protocol.

E2D08 Which of the following types of packet frames is used to transmit APRS beacon data?

A. Connect frames
B. Disconnect frames
C. Acknowledgement frames
D. Unnumbered Information frames

D To simplify the communication between stations, an APRS beacon is transmitted as an unnumbered information (UI) frame. The stations in an APRS network are not "connected" in the normal packet radio sense. The beacon frames are not directed to a specific station, and receiving stations do not acknowledge correct receipt of the frames.

E2D09 Under clear communications conditions, which of these digital communications modes has the fastest data throughput?

A. AMTOR
B. 170-Hz shift, 45 baud RTTY
C. PSK31
D. 300-baud packet

D For fastest throughput under clear communications conditions (assumed to be error-free transmission), the highest data rate is needed. Of the choices given, 300-baud packet has the highest data rate and, therefore, the fastest data throughput of these choices.

E2D10 How can an APRS station be used to help support a public service communications activity?

A. An APRS station with an emergency medical technician can automatically transmit medical data to the nearest hospital
B. APRS stations with General Personnel Scanners can automatically relay the participant numbers and time as they pass the check points
C. An APRS station with a GPS unit can automatically transmit information to show a mobile station's position during the event
D. All of these choices are correct

C At this point, it is good to remember that APRS stands for Automatic Position Reporting System. Positions of fixed stations can be entered directly into software. Mobile APRS stations equipped with a GPS (Global Positioning System) unit can automatically transmit information to show the station's position. This capability can be useful in support of a public service communications activity, such as a walk-a-thon.

E2D11 Which of the following data sources are needed to accurately transmit your geographical location over the APRS network?

A. The NMEA-0183 formatted data from a Global Positioning System (GPS) satellite receiver
B. The latitude and longitude of your location, preferably in degrees, minutes and seconds, entered into the APRS computer software
C. The NMEA-0183 formatted data from a LORAN navigation system
D. Any of these choices is correct

D Each of the statements is correct. Many TNCs have a serial port that can be connected directly to the data output jack of the GPS receiver. Software in the TNC recognizes the NMEA-0183 format and automatically extracts the latitude and longitude information for transmission over the APRS network. If the GPS receiver is not connected directly to the TNC, an operator can enter the position information by using APRS software to control the TNC.

E2E   Operating methods, part 3: operating HF digital modes; error correction

E2E01   What is a common method of transmitting data emissions below 30 MHz?
A. DTMF tones modulating an FM signal
B. FSK/AFSK
C. Pulse modulation
D. Spread spectrum

B   The most common method of transmitting data emissions below 30 MHz is by FSK (frequency-shift keying) of an RF carrier using an SSB transceiver.

E2E02   What do the letters FEC mean as they relate to digital operation?
A. Forward Error Correction
B. First Error Correction
C. Fatal Error Correction
D. Final Error Correction

A   The telecommunications industry uses FEC as a standard acronym for Forward Error Correction. FEC involves the use of sending redundant data or special codes with the data so that the receiving system can compensate for errors that occur during transmission and reception.

E2E03   How is Forward Error Correction implemented?
A. By the receiving station repeating each block of three data characters
B. By transmitting a special algorithm to the receiving station along with the data characters
C. By transmitting extra data that may be used to detect and correct transmission errors
D. By varying the frequency shift of the transmitted signal according to a predefined algorithm

C   This is a general question about Forward Error Correction systems. These systems do not require receiving stations to transmit an acknowledgement to the sending station. FEC is implemented by transmitting extra data that may be used to detect and correct transmission errors.

E2E04 What is indicated when one of the ellipses in an FSK crossed-ellipse display suddenly disappears?

A. Selective fading has occurred
B. One of the signal filters has saturated
C. The receiver has drifted 5 kHz from the desired receive frequency
D. The mark and space signal have been inverted

A A common tuning aid for digital FSK signals uses a crossed pair of ellipses. Each ellipse represents one of the two tones being sent. With the propagation effect known as selective fading, a small region of the radio spectrum undergoes a deep fade as if a narrow filter has been applied to it. This is what can make one of the ellipses on the terminal unit display disappear.

E2E05 How does ARQ accomplish error correction?

A. Special binary codes provide automatic correction
B. Special polynomial codes provide automatic correction
C. If errors are detected, redundant data is substituted
D. If errors are detected, a retransmission is requested

D ARQ stands for Automatic Repeat Request. When an error is detected in received data, the ARQ receive system sends a NAK (Not Acknowledge) message back to the transmitting system so that the data is sent again.

E2E06 What is the most common data rate used for HF packet communications?

A. 48 baud
B. 110 baud
C. 300 baud
D. 1200 baud

C The most common data rate used for HF packet communications is 300 bauds. Slower signaling rates are used for RTTY and higher rates are not permitted by regulation because they require more bandwidth.

E2E07 What is the typical bandwidth of a properly modulated MFSK16 signal?

A. 31 Hz
B. 316 Hz
C. 550 Hz
D. 2 kHz

B MFSK16 uses 16 tones, which are sent one at a time at 15.625 bauds. Because the 16 tones are spaced 15.625 Hz apart, a properly modulated MFSK16 signal has a bandwidth of 316 Hz.

E2E08 Which of the following HF digital modes can be used to transfer binary files?

A. Hellschreiber
B. PACTOR
C. RTTY
D. AMTOR

B PACTOR was developed to combine the best features of packet radio and AMTOR. Since PACTOR carries binary data, you can use this mode to transfer binary data files directly between stations. The other mode choices are not suitable for this purpose.

E2E09 Which of the following HF digital modes uses variable-length coding for bandwidth efficiency?

A. RTTY
B. PACTOR
C. MT63
D. PSK31

D Peter Martinez, G3PLX, invented Varicode as a way to improve the data rate of PSK31. In Varicode, the more common characters have shorter codes, just like in Morse code.

E2E10 This question has been removed by the Question Pool Committee.

E2E11 What is the Baudot code?

A. A code used to transmit data only in modern computer-based data systems using seven data bits
B. A binary code consisting of eight data bits
C. An alternate name for Morse code
D. The International Telegraph Alphabet Number 2 (ITA2) which uses five data bits

D Traditional RTTY communication uses the Baudot code. This digital code, also called the International Telegraph Alphabet Number 2 (ITA2), uses five data bits to represent each character, with additional start and stop bits to indicate the beginning and end of each character.

E2E12 Which of these digital communications modes has the narrowest bandwidth?

A. MFSK16
B. 170-Hz shift, 45 baud RTTY
C. PSK31
D. 300-baud packet

C PSK31 uses a data rate of 31.25 bauds. This results in an RF signal that has a bandwidth of only 31.25 Hz! This is by far the narrowest bandwidth of any digital communications mode.

Radio Wave Propagation

There will be three questions on your Extra class examination from the Radio Wave Propagation subelement. These three questions will be taken from the three groups of questions labeled E3A through E3C.

E3A Propagation and technique, part 1: Earth-Moon-Earth communications (EME); meteor scatter

E3A01 What is the approximate maximum separation along the surface of the Earth between two stations communicating by moonbounce?

A. 500 miles if the moon is at perigee
B. 2000 miles, if the moon is at apogee
C. 5000 miles, if the moon is at perigee
D. 12,000 miles, as long as both can "see" the moon

D Two stations must be able to simultaneously see the moon in order to communicate by reflecting VHF or UHF signals off the lunar surface. Those stations may be separated by nearly 180° of arc on the Earth's surface — a distance of more than 11,000 miles. There is no specific maximum distance between two stations to communicate via moonbounce, as long as they have a mutual lunar window. In other words, the moon must be above the radio horizon where both stations can "see" it at the same time.

E3A02 What characterizes libration fading of an earth-moon-earth signal?

A. A slow change in the pitch of the CW signal
B. A fluttery irregular fading
C. A gradual loss of signal as the sun rises
D. The returning echo is several Hertz lower in frequency than the transmitted signal

B Libration fading is multipath scattering of the radio waves from the very large (2000 mile diameter) and rough Moon surface combined with the relatively short-term variations of the Moon in its orbit. Libration fading of an EME signal is characterized in general as fluttery, rapid, irregular fading not unlike that observed in tropospheric-scatter propagation. Fading can be very deep, 20 dB or more, and the maximum fading will depend on the operating frequency. You can see this in the graph in Figure E3A02.

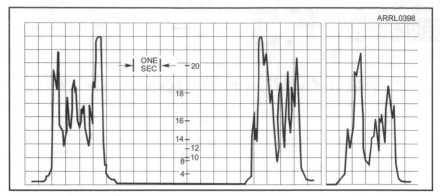

Figure E3A02 — This graph is a chart recording that shows the deep libration fading of echoes from the moon's surface received at W2NFA

E3A03   When scheduling EME contacts, which of these conditions will generally result in the least path loss?

A. When the moon is at perigee
B. When the moon is full
C. When the moon is at apogee
D. When the MUF is above 30 MHz

A   The moon's orbit is slightly elliptical, with the closest distance (perigee) being 225,000 miles and the furthest (apogee) being 252,000 miles. EME path loss is typically 2 dB less at perigee.

E3A04   What type of receiving system is desirable for EME communications?

A. Equipment with very wide bandwidth
B. Equipment with very low dynamic range
C. Equipment with very low gain
D. Equipment with very low noise figures

D   A low-noise receiving setup is essential for successful EME work. Since many of the signals to be copied on EME are barely, but not always, out of the noise, a low-noise receiver is required. At 144 MHz a noise figure of under 0.5 dB will make cosmic noise the limiting factor of what you're able to hear. This can be achieved with commonly available equipment. At UHF, you'll want the lowest noise figure you can attain. With GaAsFET devices, you should be able to buy or build a preamplifier with a 0.5 dB noise figure.

E3A05 What transmit and receive time sequencing is normally used on 144 MHz when attempting an EME contact?

A. Two-minute sequences, where one station transmits for a full two minutes and then receives for the following two minutes
B. One-minute sequences, where one station transmits for one minute and then receives for the following one minute
C. Two-and-one-half minute sequences, where one station transmits for a full 2.5 minutes and then receives for the following 2.5 minutes
D. Five-minute sequences, where one station transmits for five minutes and then receives for the following five minutes

A For 144-MHz contacts, a two-minute calling sequence is used. The eastern-most station transmits first for two full minutes, and then that station receives for two full minutes while the western-most station transmits.

E3A06 What transmit and receive time sequencing is normally used on 432 MHz when attempting an EME contact?

A. Two-minute sequences, where one station transmits for a full two minutes and then receives for the following two minutes
B. One-minute sequences, where one station transmits for one minute and then receives for the following one minute
C. Two-and-one-half minute sequences, where one station transmits for a full 2.5 minutes and then receives for the following 2.5 minutes
D. Five-minute sequences, where one station transmits for five minutes and then receives for the following five minutes

C For UHF EME contacts, operators use 2½ minute calling sequences. The eastern-most station transmits first, for 2½ minutes. Then the eastern station listens for the next 2½ minutes, while the western-most station transmits.

E3A07 What frequency range would you normally tune to find EME stations in the 2 meter band?

A. 144.000 - 144.001 MHz
B. 144.000 - 144.100 MHz
C. 144.100 - 144.300 MHz
D. 145.000 - 145.100 MHz

B Most EME contacts are made in the weak-signal portion of the bands. On 2 meters, this is 144.000 to 144.100 MHz. EME contacts are generally made by prearranged schedule, although some contacts are made at random.

E3A08 What frequency range would you normally tune to find EME stations in the 70 cm band?

A. 430.000 - 430.150 MHz
B. 430.100 - 431.100 MHz
C. 431.100 - 431.200 MHz
D. 432.000 - 432.100 MHz

D As on 2 meters, EME contacts are made in the weak-signal portion of the band. That means 432.000 to 432.100 MHz.

E3A09 When a meteor strikes the Earth's atmosphere, a cylindrical region of free electrons is formed at what layer of the ionosphere?

A. The E layer
B. The F1 layer
C. The F2 layer
D. The D layer

A Meteor trails are formed at approximately the altitude of the ionospheric E layer, 50 to 75 miles above the Earth. This is illustrated in Figure E3A09. As a meteoroid speeds through the upper atmosphere, it begins to burn or vaporize as it collides with air molecules. This action creates heat and light and leaves a trail of free electrons and positively charged ions behind as the meteoroid races along. Trail size is directly dependent on size and speed. A typical meteoroid is the size of a grain of sand. A particle this size creates a trail about 3 feet in diameter and 12 miles or longer, depending on speed.

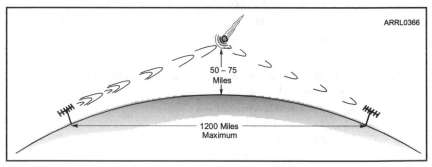

Figure E3A09 — Meteor-scatter communication makes extended-range VHF contacts possible by using the ionized meteor trail as a reflector 50-75 miles above the Earth.

E3A10 Which range of frequencies is well suited for meteor-scatter communications?

A. 1.8 - 1.9 MHz
B. 10 - 14 MHz
C. 28 - 148 MHz
D. 220 - 450 MHz

C The electron density in a typical meteor trail will strongly affect radio waves between 28 and 148 MHz. Signal frequencies as low as 20 MHz and as high as 432 MHz will be usable for meteor-scatter communication at times.

E3A11 What transmit and receive time sequencing is normally used on 144 MHz when attempting a meteor-scatter contact?

A. Two-minute sequences, where one station transmits for a full two minutes and then receives for the following two minutes
B. One-minute sequences, where one station transmits for one minute and then receives for the following one minute
C. 15-second sequences, where one station transmits for 15 seconds and then receives for the following 15 seconds
D. 30-second sequences, where one station transmits for 30 seconds and then receives for the following 30 seconds

C The secret to successful meteor-scatter communication is short transmissions. The accepted convention for timing transmissions breaks each minute into four 15-second periods. The station at the western end of the path transmits during the first and third period of each minute. During the second and fourth 15-second periods, the eastern station transmits.

E3B **Propagation and technique, part 2: transequatorial; long path; gray line; multi-path propagation**

E3B01 What is transequatorial propagation?

A. Propagation between two points at approximately the same distance north and south of the magnetic equator
B. Propagation between any two points located on the magnetic equator
C. Propagation between two continents by way of ducts along the magnetic equator
D. Propagation between two stations at the same latitude

A Transequatorial propagation (TE) is a form of F-layer ionospheric propagation that was discovered by amateurs. TE allows hams on either side of the magnetic equator to communicate with each other. As the signal frequency increases, TE propagation becomes more restricted to regions equidistant from, and perpendicular to, the magnetic equator.

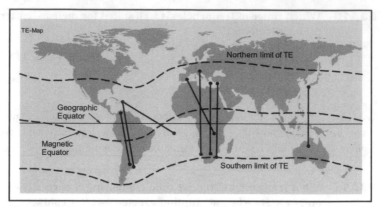

Figure E3B01 — This world map shows TE paths worked by amateurs on 144 MHz. Notice the symmetrical distribution of stations with respect to the magnetic equator. Because the poles of the Earth's magnetic field are not aligned with its geomagnetic axis, the magnetic equator does not follow the geographic equator and is somewhat tilted and distorted.

E3B02　What is the approximate maximum range for signals using transequatorial propagation?

A. 1000 miles
B. 2500 miles
C. 5000 miles
D. 7500 miles

C　Maximum TE range is approximately 5000 miles — 2500 miles on each side of the magnetic equator.

E3B03　What is the best time of day for transequatorial propagation?

A. Morning
B. Noon
C. Afternoon or early evening
D. Late at night

C　Ionization levels that support TE are forming during the morning, are well established by noon and may last until after midnight. The best (peak) time is in the afternoon and early evening hours.

E3B04 What type of propagation is probably occurring if an HF beam antenna must be pointed in a direction 180 degrees away from a station to receive the strongest signals?

A. Long-path
B. Sporadic-E
C. Transequatorial
D. Auroral

A Propagation between any two points on the Earth's surface is usually by the shortest direct route, which is a great-circle path between the two points. A great circle is an imaginary line, or ring, drawn around the Earth, where a plane passing through the center of the Earth would intersect the Earth's surface. Unless the two points are opposite each other, one way will be shorter than the other will, and this is the usual propagation path. Under certain conditions, propagation may favor the longer path, traveling "the other way." Under those conditions, you'll have to point a beam antenna in the direction of the long path.

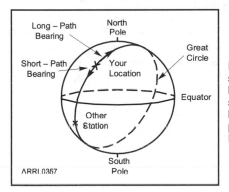

Figure E3B04 — A drawing showing a great circle path between two stations. The short-path and long-path bearings are shown from the perspective of the Northern Hemisphere station.

E3B05 Which amateur bands typically support long-path propagation?

A. 160 to 40 meters
B. 30 to 10 meters
C. 160 to 10 meters
D. 6 meters to 2 meters

C Long-path propagation can occur on any band that provides ionospheric propagation. That means you might experience long-path propagation on the 160 to 10-meter bands. Long-path propagation has been observed on 6 meters, but is quite uncommon.

E3B06   Which of the following amateur bands most frequently provides long-path propagation?

A.   80 meters
B.   20 meters
C.   10 meters
D.   6 meters

B   You can consistently make use of long-path enhancement on the 20-meter band. All it takes is a modest beam antenna with a relatively high gain compared to a dipole, such as a Yagi or quad, at a height that enhances radiation at low vertical angles.

E3B07   Which of the following could account for hearing an echo on the received signal of a distant station?

A.   High D layer absorption
B.   Meteor scatter
C.   Transmit frequency is higher than the MUF
D.   Receipt of a signal by more than one path

D   If you are in North America and hear an echo on signals from European stations when your antenna is pointing toward Europe, the echo may be coming in by long-path propagation. Because the signals have to travel much further on the long path, they will be delayed compared to the short-path signals.

E3B08   What type of propagation is probably occurring if radio signals travel along the terminator between daylight and darkness?

A.   Transequatorial
B.   Sporadic-E
C.   Long-path
D.   Gray-line

D   The gray line is a band around the Earth that separates daylight from darkness. Astronomers call this line the terminator. Propagation along the gray line can be very efficient because absorption by the D-layer is often minimized near the terminator. (Also see the discussion for E3B10.)

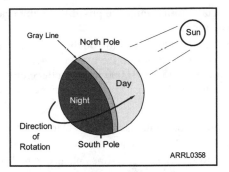

Figure E3B08 — The gray line is a transition region between daylight and darkness. One side of the Earth is coming into sunrise and the other is just past sunset.

E3B09 At what time of day is gray-line propagation most prevalent?

A. At sunrise and sunset
B. When the sun is directly above the location of the transmitting station
C. When the sun is directly overhead at the middle of the communications path between the two stations
D. When the sun is directly above the location of the receiving station

A Look for gray-line propagation at twilight, around sunrise and sunset. Point your antenna along the terminator in either direction.

E3B10 What is the cause of gray-line propagation?

A. At midday, the sun, being directly overhead, superheats the ionosphere causing increased refraction of radio waves
B. At twilight, solar absorption drops greatly, while atmospheric ionization is not weakened enough to reduce the MUF
C. At darkness, solar absorption drops greatly, while atmospheric ionization remains steady
D. At mid afternoon, the sun heats the ionosphere, increasing radio wave refraction and the MUF

B Propagation along the gray line can be very efficient. The major reason is that the D-layer, which absorbs HF signals, disappears rapidly on the sunset side of the gray line, and has yet to build up on the sunrise side. By contrast, the much higher F-layer forms earlier and lasts much longer.

E3B11 What communications are possible during gray-line propagation?

A. Contacts up to 2,000 miles only on the 10-meter band
B. Contacts up to 750 miles on the 6- and 2-meter bands
C. Contacts up to 8,000 to 10,000 miles on three or four HF bands
D. Contacts up to 12,000 to 15,000 miles on the 2 meter and 70 centimeter bands

C Gray-line propagation contacts up to 8,000 to 10,000 miles are possible. The three or four lowest-frequency amateur bands (160, 80, 40 and 30 meters) are the most likely to experience gray-line enhancement, because they are the most affected by D-layer absorption.

E3C Propagation and technique, part 3: Auroral propagation; selective fading; radio-path horizon; take-off angle over flat or sloping terrain; earth effects on propagation; less common propagation modes

E3C01 What effect does auroral activity have on radio communications?

A. Signals experience long-delay echo
B. FM communications are clearer
C. CW signals have a clearer tone
D. CW signals have a fluttery tone

D Auroral propagation occurs when VHF radio waves are reflected from ionization associated with an auroral curtain. The reflecting properties of an aurora vary rapidly, so signals received via this mode are badly distorted by multipath effects. CW is the most effective mode for auroral work. The CW signal is so badly distorted that it is most often a buzzing or hissing sound rather than a pure tone.

E3C02 **What is the cause of auroral activity?**
  A. Reflections in the solar wind
  B. A low sunspot level
  C. The emission of charged particles from the sun
  D. Meteor showers concentrated in the northern latitudes

C Aurora results from a large-scale interaction between the magnetic field of the Earth and electrically charged particles arriving from the Sun. During times of enhanced solar activity, electrically charged particles are ejected from the surface of the Sun. These particles form a solar wind, which travels through space. When the solar wind interacts with the Earth's magnetic field, its charged particles may cause an aurora.

E3C03 **Where in the ionosphere does auroral activity occur?**
  A. At F-region height
  B. In the equatorial band
  C. At D-region height
  D. At E-region height

D Auroral activity is caused by ionization at an altitude of about 70 miles above Earth. This is very near the altitude (height) of the ionospheric E layer.

E3C04 **Which emission mode is best for auroral propagation?**
  A. CW
  B. SSB
  C. FM
  D. RTTY

A Signals received by auroral propagation are badly distorted because of the erratic nature of reflection from the auroral region. For that reason, CW is the most effective mode for auroral work. While SSB may be usable at 6 meters when signals are strong and the operator speaks slowly and distinctly, it is rarely usable at 2 meters or higher frequencies.

E3C05 **What causes selective fading?**
  A. Small changes in beam heading at the receiving station
  B. Phase differences in the received signal caused by different paths
  C. Large changes in the height of the ionosphere
  D. Time differences between the receiving and transmitting stations

B Selective fading occurs because of phase differences between radio-wave components of the same transmission, as experienced at the receiving station. The result is distortion that may range from mild to severe. It is possible for components of the same signal that are only a few kilohertz apart (such as a carrier and the sidebands in an AM signal) to be acted upon differently by the ionosphere. This causes the modulation sidebands to arrive at the receiver out of phase.

E3C06 How much farther does the VHF/UHF radio-path horizon distance exceed the geometric horizon?

A. By approximately 15% of the distance
B. By approximately twice the distance
C. By approximately one-half the distance
D. By approximately four times the distance

A The radio horizon is approximately 15% farther than the geometric horizon. Under normal conditions, the structure of the atmosphere near the Earth causes radio waves to bend into a curved path that keeps them nearer to the Earth than would be the case for true straight-line travel.

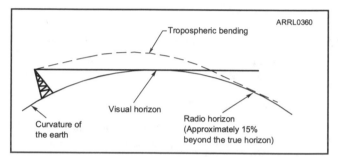

Figure E3C06 — Under normal conditions, bending in the troposphere causes VHF and UHF radio waves to be returned to Earth beyond the visible horizon.

E3C07 How does the radiation pattern of a 3-element, horizontally polarized beam antenna vary with height above ground?

A. The main lobe takeoff angle increases with increasing height
B. The main lobe takeoff angle decreases with increasing height
C. The horizontal beam width increases with height
D. The horizontal beam width decreases with height

B In general, the radiation takeoff angle from a Yagi antenna with horizontally mounted elements decreases as the antenna height increases above flat ground. So, if you raise the height of your antenna, the takeoff angle will decrease.

E3C08   What is the name of the high-angle wave in HF propagation that travels for some distance within the F2 region?

A. Oblique angle ray
B. Pedersen ray
C. Ordinary ray
D. Heaviside ray

B   Radio waves may at times propagate for some distance through the F region of the ionosphere before being bent back to the Earth or exiting the ionosphere to space. Studies have shown that a signal radiated at a medium elevation angle sometimes reaches the Earth at a greater distance than a lower-angle wave. This higher-angle wave, called the Pedersen ray, is believed to penetrate the F region farther than lower-angle rays. In the less densely ionized upper edge of the region, the amount of refraction is less, nearly equaling the curvature of the region itself as it encircles the Earth. Figure E3C08 shows how the Pedersen ray could provide propagation beyond the normal single-hop distance.

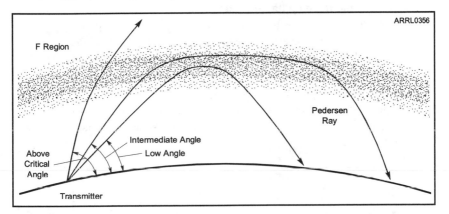

Figure E3C08 — This diagram shows a radio wave entering the F region at an intermediate angle and penetrating higher than normal into the F region. This wave, called the Pedersen ray, then travels within that region for some distance before being bent enough to return to Earth.

E3C09 What effect is usually responsible for propagating a VHF signal over 500 miles?

A. D-region absorption
B. Faraday rotation
C. Tropospheric ducting
D. Moonbounce

C VHF propagation is usually limited to distances of approximately 500 miles. At times, however, VHF communications are possible with stations up to 2000 or more miles away. Certain weather conditions cause duct-like structures to form in the troposphere, simulating propagation within a waveguide. Such ducts cause VHF radio waves to follow the curvature of the Earth for hundreds, or thousands, of miles. This form of propagation is called tropospheric ducting.

E3C10 How does the performance of a horizontally polarized antenna mounted on the side of a hill compare with the same antenna mounted on flat ground?

A. The main lobe takeoff angle increases in the downhill direction
B. The main lobe takeoff angle decreases in the downhill direction
C. The horizontal beam width decreases in the downhill direction
D. The horizontal beam width increases in the uphill direction

B A horizontal Yagi antenna installed above a slope will have a lower takeoff angle in the downward direction of the slope than a similar Yagi mounted above flat Earth. The steeper the slope, the lower the takeoff angle will be.

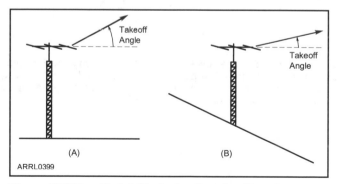

Figure E3C10 — Part A illustrates the takeoff angle for radio waves leaving a Yagi antenna with horizontal elements over flat ground. Higher antenna elevations result in smaller takeoff angles (see the discussion for E3C07). Part B shows the takeoff angle for a similar antenna over sloping ground. For steeper slopes away from the front of the antenna, the takeoff angle is lowered.

E3C11 From the contiguous 48 states, in which approximate direction should an antenna be pointed to take maximum advantage of auroral propagation?

A. South
B. North
C. East
D. West

B Auroras occur around the magnetic poles. In the northern hemisphere, stations point their antennas north and "bounce" their signals off the auroral zone. Operators should move their antennas to find the bearing at which the reflection is strongest.

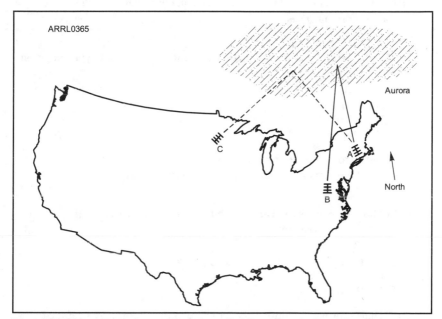

Figure E3C11 — To make contacts by reflecting signals from the aurora, stations in the Northern Hemisphere point their antennas toward the North Pole — or toward the North Magnetic Pole if they are close enough for there to be a significant difference in bearings. Aim your antenna in different directions to find strongest reflection.

E3C12 As the frequency of a signal is increased, how does its ground wave propagation change?

A. It increases
B. It decreases
C. It stays the same
D. Radio waves don't propagate along the Earth's surface

B   Ground-wave propagation refers to diffraction of vertically polarized waves. Ground-wave propagation is most noticeable on the AM broadcast band and the 160 and 80 meter amateur bands. Practical ground-wave communications distances on these bands often extend to 120 miles or more. Ground-wave loss increases significantly with higher frequencies, so it is not useful even at 40 meters. Although the term ground-wave propagation is often applied to any short-distance communication, the actual mechanism is unique to the lower frequencies.

E3C13 What type of polarization does most ground-wave propagation have?

A. Vertical
B. Horizontal
C. Circular
D. Elliptical

A   Most ground-wave propagation uses vertical polarization. Signals with horizontal polarization do not propagate by ground-wave.

E3C14 Why does the radio-path horizon distance exceed the geometric horizon?

A. E-region skip
B. D-region skip
C. Auroral skip
D. Radio waves may be bent

D   Under normal conditions, the structure of the atmosphere near the Earth causes radio waves to bend into a curved path. That effect keeps the radio waves nearer to the Earth than true straight-line travel would. (See the discussion for E3C06.)

Amateur Radio Practices

There will be five questions on your Extra Class examination from the Amateur Radio Practices subelement. These five questions will be taken from the five groups of questions labeled E4A through E4E.

E4A Test equipment: analog and digital instruments; spectrum and network analyzers, antenna analyzers; oscilloscopes; testing transistors; RF measurements

E4A01 How does a spectrum analyzer differ from a conventional oscilloscope?

A. A spectrum analyzer measures ionospheric reflection; an oscilloscope displays electrical signals

B. A spectrum analyzer displays the peak amplitude of signals; an oscilloscope displays the average amplitude of signals

C. A spectrum analyzer displays signals in the frequency domain; an oscilloscope displays signals in the time domain

D. A spectrum analyzer displays radio frequencies; an oscilloscope displays audio frequencies

C Use a spectrum analyzer to view signals in the frequency domain (amplitude vs frequency) and an oscilloscope to view them in the time domain (amplitude vs time). For both instruments the vertical axis is signal amplitude. The difference is that the spectrum analyzer displays frequency along the horizontal axis and the oscilloscope displays time.

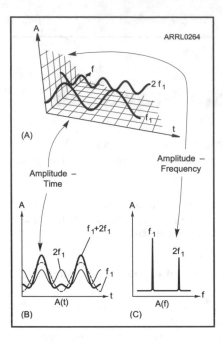

Figure E4A01 — This diagram shows how the complex signal at A can be viewed in the time domain by an oscilloscope or in the frequency domain by a spectrum analyzer. The oscilloscope would show the sum of the two signals as at B. The spectrum analyzer would show two separate components at different frequencies as at C. This type of display is usually more useful in analyzing complex signals.

E4A02 Which of the following parameters would a typical spectrum analyzer display on the horizontal axis?

A. SWR
B. Q
C. Time
D. Frequency

D The spectrum analyzer shows frequency on the horizontal axis.

E4A03 Which of the following parameters would a typical spectrum analyzer display on the vertical axis?

A. Amplitude
B. Duration
C. SWR
D. Q

A The vertical axis of the spectrum analyzer shows signal amplitude.

E4A04 Which of the following test instruments is used to display spurious signals from a radio transmitter?

A. A spectrum analyzer
B. A wattmeter
C. A logic analyzer
D. A time-domain reflectometer

A A spectrum analyzer is the best instrument for displaying spurious signals because they can be identified by frequency. The other instruments listed here do not display signal frequency and so are not suited for this purpose.

E4A05 Which of the following test instruments is used to display intermodulation distortion products in an SSB transmission?

A. A wattmeter
B. A spectrum analyzer
C. A logic analyzer
D. A time-domain reflectometer

B This question is similar to the previous one in that a frequency domain measurement is required, so the spectrum analyzer is the correct choice.

E4A06 Which of the following could be determined with a spectrum analyzer?

A. The degree of isolation between the input and output ports of a 2 meter duplexer
B. Whether a crystal is operating on its fundamental or overtone frequency
C. The spectral output of a transmitter
D. All of these choices are correct

D All of the choices are measurements of amplitude (attenuation between isolator ports is an amplitude measurement) versus frequency, so all can be measured with a spectrum analyzer.

E4A07 Which of the following is an advantage of using an antenna analyzer vs. a SWR bridge to measure antenna SWR?

A. Antenna analyzers automatically tune your antenna for resonance
B. Antenna analyzers typically do not need an external RF source
C. Antenna analyzers typically display a time-varying representation of the modulation envelope
D. All of these answers are correct

B An antenna analyzer has a variable frequency source built-in, so you don't have to key or adjust a transmitter to measure SWR across an entire band. Antenna analyzers also provide information about the antenna's feed point impedance that is unavailable from an SWR bridge.

E4A08 Which of the following instruments would be best for measuring the SWR of a beam antenna?

A. A spectrum analyzer
B. A Q meter
C. An ohmmeter
D. An antenna analyzer

D See the discussion for E4A07.

E4A09 Which of the following is most important when adjusting PSK31 transmitting levels?

A. Power output
B. PA current
C. ALC level
D. SWR

C If the level of the audio PSK31 signal into the transmitter is too high, it will overdrive the power amplifier stages, producing splatter and harmonics. The transmitter ALC (automatic level control) circuit limits the amount of input power to the final amplifier so that it is not overdriven. By watching the ALC control voltage, the PSK31 input level can be adjusted to an appropriate level that avoids overdrive.

E4A10   Which of the following is a useful test for a functioning NPN transistor in an active circuit where the transistor should be biased "on"?

A. Measure base-to-emitter resistance with an ohmmeter; it should be approximately 6 to 7 ohms

B. Measure base-to-emitter resistance with an ohmmeter; it should be approximately 0.6 to 0.7 ohms

C. Measure base-to-emitter voltage with a voltmeter; it should be approximately 6 to 7 volts

D. Measure base-to-emitter voltage with a voltmeter; it should be approximately 0.6 to 0.7 volts

D   When an NPN transistor is biased on, base-to-emitter voltage (V_{BE}) should measure 0.6 to 0.75 V from base to emitter with the positive voltmeter lead connected to the base.

E4A11   Which of the following test instruments can be used to indicate pulse conditions in a digital logic circuit?

A. A logic probe

B. An ohmmeter

C. An electroscope

D. A Wheatstone bridge

A   A logic probe — a specialized type of voltmeter for measuring digital signals — can indicate high or low signal levels. Most logic probes can also display the presence of short pulses and pulse waveforms.

E4A12   Which of the following procedures is an important precaution to follow when connecting a spectrum analyzer to a transmitter output?

A. Use high quality double shielded coaxial cables to reduce signal losses

B. Attenuate the transmitter output going to the spectrum analyzer

C. Match the antenna to the load

D. All of these choices are correct

B   Most spectrum analyzer inputs are at risk of damage from input signals stronger than 1 watt. Therefore, a signal coming from the transmitter needs to pass through an attenuator before it is applied to the spectrum analyzer input.

E4B Measurement technique and limitations: instrument accuracy and performance limitations; probes; techniques to minimize errors; measurement of "Q"; instrument calibration

E4B01 Which of the following is a characteristic of a good harmonic frequency marker?

A. Wide tuning range
B. Frequency stability
C. Linear output amplifier
D. All of the above

B Frequency markers are used for calibrating receivers, so frequency stability is the most important characteristic of a marker generator.

E4B02 Which of the following factors most affects the accuracy of a frequency counter?

A. Input attenuator accuracy
B. Time base accuracy
C. Decade divider accuracy
D. Temperature coefficient of the logic

B A frequency counter counts the number of pulses applied to its input during a period of time and displays the results. The time base determines the period for counting pulses, so the accuracy of the time base determines the accuracy of the counter. The stability of the time base determines the stability of the frequency counter.

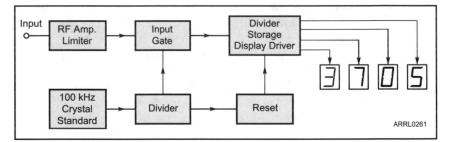

Figure E4B02 — The block diagram of the basic parts of a frequency counter. The time base is based on the output of a highly stable and accurate crystal oscillator.

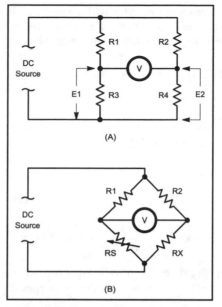

Figure E4B03 — A Wheatstone bridge circuit. A bridge circuit is actually a pair of voltage dividers (A) with R1 = R2. One of the dividers is adjustable as in (B). When the unknown impedance is attached at RX and the bridge adjusted so that the voltages E1 and E2 are equal, the voltmeter V indicates a null or zero voltage. The value of RS is then equal to RX.

E4B03 What is an advantage of using a bridge circuit to measure impedance?

A. It provides an excellent match under all conditions
B. It is relatively immune to drift in the signal generator source
C. The measurement is based on obtaining a null in voltage, which can be done very precisely
D. It can display results directly in Smith chart format

C A bridge circuit's null is very sharp so that it is easy to adjust the bridge precisely to the deepest point of the null. That means the unknown impedance value is also measured precisely.

E4B04   If a frequency counter with a specified accuracy of +/– 1.0 ppm reads 146,520,000 Hz, what is the most the actual frequency being measured could differ from the reading?

A.   165.2 Hz
B.   14.652 kHz
C.   146.52 Hz
D.   1.4652 MHz

C   Here you have the first of a series of questions for which the answer is computed by a process based on the formula

Error = Frequency × Accuracy

In this case, when you substitute the numbers, you get

$$\text{Error} = 146,520,000 \text{ Hz} \times \frac{\pm 1}{1,000,000} = \pm 146.520 \text{ Hz}$$

It makes the math a bit easier to realize that if you write the frequency in MHz then it will cancel with the accuracy in ppm. Try that on the next problem.

E4B05   If a frequency counter with a specified accuracy of +/– 0.1 ppm reads 146,520,000 Hz, what is the most the actual frequency being measured could differ from the reading?

A.   14.652 Hz
B.   0.1 MHz
C.   1.4652 Hz
D.   1.4652 kHz

A   Use the same formula as you did for the previous question, but use the shortcut this time. You find

Error (in Hz) = Frequency (in MHz) × Accuracy (in ppm)

= 146.52 × ±0.1 = ±14.652 Hz

That was a bit easier, wasn't it?

E4B06   If a frequency counter with a specified accuracy of +/–10 ppm reads 146,520,000 Hz, what is the most the actual frequency being measured could differ from the reading?

A.   146.52 Hz
B.   10 Hz
C.   146.52 kHz
D.   1465.20 Hz

D   Use the same formula and you'll find

Error = 146.52 MHz × ±10 ppm = ±1465.20 Hz

E4B07 How much power is being absorbed by the load when a directional power meter connected between a transmitter and a terminating load reads 100 watts forward power and 25 watts reflected power?

A. 100 watts
B. 125 watts
C. 25 watts
D. 75 watts

D A directional wattmeter reads the total amount of power in a feed line traveling in each direction. The actual amount of power being absorbed by the load is the difference between the forward and reflected power readings, Forward power − Reflected power = 100 − 25 = 75 watts.

E4B08 Which of the following is good practice when using an oscilloscope probe?

A. Keep the ground connection of the probe as short as possible
B. Never use a high impedance probe to measure a low impedance circuit
C. Never use a DC-coupled probe to measure an AC circuit
D. All of these choices are correct

A High-frequency signals being measured with an oscilloscope probe must flow through the probe tip, to the 'scope's input circuit through a short coaxial cable, and back to the circuit through the ground connection lead. The probe tip connection and coaxial cable have very good high-frequency characteristics, but an excessively long ground connection introduces inductance into the signal path that can cause distortion or high-frequency rolloff of the signal.

E4B09 Which of the following is a characteristic of a good DC voltmeter?

A. High reluctance input
B. Low reluctance input
C. High impedance input
D. Low impedance input

C A good voltmeter, either ac or dc, measures voltage while drawing as little current as possible from the circuit being measured. This requires the meter to have a high impedance input circuit.

E4B10 What is indicated if the current reading on an RF ammeter placed in series with the antenna feedline of a transmitter increases as the transmitter is tuned to resonance?

A. There is possibly a short to ground in the feedline
B. The transmitter is not properly neutralized
C. There is an impedance mismatch between the antenna and feedline
D. There is more power going into the antenna

D Higher feed line current means that more power is flowing to the antenna.

E4B11 Which of the following describes a method to measure intermodulation distortion in an SSB transmitter?

A. Modulate the transmitter with two non-harmonically related radio frequencies and observe the RF output with a spectrum analyzer
B. Modulate the transmitter with two non-harmonically related audio frequencies and observe the RF output with a spectrum analyzer
C. Modulate the transmitter with two harmonically related audio frequencies and observe the RF output with a peak reading wattmeter
D. Modulate the transmitter with two harmonically related audio frequencies and observe the RF output with a logic analyzer

B As described in question E4A04, a spectrum analyzer is the best instrument to use to look for spurious transmitter outputs. The test tones into the transmitter should not be harmonically related (integer multiples of the same frequency) so that any spurious signals do not occur at the same frequency. Spurious outputs from non-harmonically related signals will appear as independent signal components.

E4B12 How should a portable SWR analyzer be connected when measuring antenna resonance and feedpoint impedance?

A. Loosely couple the analyzer near the antenna base
B. Connect the analyzer via a high-impedance transformer to the antenna
C. Connect the antenna and a dummy load to the analyzer
D. Connect the antenna feed line directly to the analyzer's connector

D The analyzer is actually a variable-frequency low-power transmitter with a built-in impedance bridge. So SWR is measured just as with standalone transmitters and SWR meters — the feed line is connected directly to the analyzer's output.

E4B13 What is the significance of voltmeter sensitivity expressed in ohms per volt?

A. The full scale reading of the voltmeter multiplied by its ohms per volt rating will provide the input impedance of the voltmeter

B. When used as a galvanometer, the reading in volts multiplied by the ohms/volt will determine the power drawn by the device under test

C. When used as an ohmmeter, the reading in ohms divided by the ohms/volt will determine the voltage applied to the circuit

D. When used as an ammeter, the full scale reading in amps divided by ohms/volt will determine the size of shunt needed

A Ohms/volt is a measure of how sensitive the meter is because it is the reciprocal of the amount of current (volts divided by ohms) required for a full-scale reading. Higher values of ohms/volt mean lower input currents for an equivalent voltage measurement and less loading of the circuit being tested.

E4B14 How is the compensation of an oscilloscope probe typically adjusted?

A. A square wave is observed and the probe is adjusted until the horizontal portions of the displayed wave is as nearly flat as possible

B. A high frequency sine wave is observed, and the probe is adjusted for maximum amplitude

C. A frequency standard is observed, and the probe is adjusted until the deflection time is accurate

D. A DC voltage standard is observed, and the probe is adjusted until the displayed voltage is accurate

A Most oscilloscopes include a circuit that outputs a square wave calibrator signal of known voltage and frequency. The probe can be connected to the calibrator output and adjusted until the flat portions of the square wave are parallel to the zero-voltage horizontal axis, not tilted or rounded in any way. This means the probe's frequency response has been adjusted properly and input signals will not be distorted.

E4B15 What happens if a dip-meter is too tightly coupled to a tuned circuit being checked?

A. Harmonics are generated

B. A less accurate reading results

C. Cross modulation occurs

D. Intermodulation distortion occurs

B When a dip meter it is too tightly coupled with the tuned circuit being checked, a less accurate reading results. Whenever two circuits are coupled, no matter how loosely, each circuit affects the other to some extent. Too tight a coupling almost certainly will yield an inaccurate reading on the dip meter.

E4B16 Which of these factors limits the accuracy of a D'Arsonval-type meter?

A. Its magnetic flux density
B. Coil impedance
C. Deflection rate
D. Electromagnet current

B Since a D'Arsonval, or moving-coil meter, is an analog meter with a mechanical movement, you should expect that the mechanical tolerance would be included as a limiting factor. Calibration and coil impedance are also important factors.

E4B17 Which of the following can be used as a relative measurement of the Q for a series-tuned circuit?

A. The inductance to capacitance ratio
B. The frequency shift
C. The bandwidth of the circuit's frequency response
D. The resonant frequency of the circuit

C The Q of a resonant circuit equals resonant frequency (f_0) divided by bandwidth (BW). By measuring the circuit's frequency response around its resonant frequency, both f_0 and BW can be determined and Q calculated.

E4C   Receiver performance characteristics, part 1: phase noise, capture effect, noise floor, image rejection, MDS, signal-to-noise-ratio; selectivity

E4C01 What is the effect of excessive phase noise in the local oscillator section of a receiver?

A. It limits the receiver ability to receive strong signals
B. It reduces the receiver sensitivity
C. It decreases the receiver third-order intermodulation distortion dynamic range
D. It can cause strong signals on nearby frequencies to interfere with reception of weak signals

D One result of receiver phase noise is that as you tune toward a strong signal, the receiver noise floor appears to increase. In other words, you hear an increasing amount of noise in an otherwise quiet receiver as you tune toward the strong signal. This means that strong signals may interfere with the reception of a nearby weak signal.

E4C02 **Which of the following is the result of the capture effect in an FM receiver?**

A. All signals on a frequency are demodulated
B. None of the signals could be heard
C. The strongest signal received is the only demodulated signal
D. The weakest signal received is the only demodulated signal

C The capture effect in FM receivers results in the loudest signal received being the only signal demodulated, even if it is only two or three times (3 to 5 dB) stronger than other signals on the same frequency. This can be an advantage if you want to receive the strong signal. However, the capture effect can prevent you from hearing a weaker signal in the presence of a stronger one.

E4C03 **What is the term for the blocking of one FM phone signal by another, stronger FM phone signal?**

A. Desensitization
B. Cross-modulation interference
C. Capture effect
D. Frequency discrimination

C See the discussion for E4C02.

E4C04 **What is meant by the noise floor of a receiver?**

A. The minimum level of noise at the audio output when the RF gain is turned all the way down
B. The equivalent phase noise power generated by the local oscillator
C. The minimum level of noise that will overload the RF amplifier stage
D. The equivalent input noise power when the antenna is replaced with a matched dummy load

D The noise floor of a receiver represents the level of a signal that equals the noise level generated within the receiver. It is the smallest level of input signal that can be detected and is determined by the receiver's internal electronics.

E4C05   What does a value of –174 dBm/Hz represent with regard to the noise floor of a receiver?

A. The minimum detectable signal as a function of receive frequency
B. The theoretical noise at the input of a perfect receiver at room temperature
C. The noise figure of a 1 Hz bandwidth receiver
D. The galactic noise contribution to minimum detectable signal

B   Noise in a receiver is primarily caused by temperature-related movement of electrons in the receiver's input circuits. The amount of noise power also depends on the receiver's bandwidth, with wider bandwidths allowing more noise power to be received. The theoretical noise floor limit is a level of –174 dBm in a bandwidth of 1 Hz.

E4C06   The thermal noise value of a receiver is –174 dBm/Hz. What is the theoretically best minimum detectable signal for a 400 Hz bandwidth receiver?

A. 174 dBm
B. –164 dBm
C. –155 dBm
D. –148 dBm

D   Since noise power is directly proportional to bandwidth, the ratio of filter bandwidths also determines the amount of additional noise power received according to the following formula:

$$\text{MDS} = -174 + 10 \log \left(\frac{\text{filter bandwidth}}{1\,\text{Hz}} \right)$$

Using the numbers in the question,

MDS = -174 dBm + 10 log (400) = -174 dBm + 26 dB = -148 dBm

E4C07   What does the MDS of a receiver represent?

A. The meter display sensitivity
B. The minimum discernible signal
C. The multiplex distortion stability
D. The maximum detectable spectrum

B   The MDS or minimum discernable signal (also minimum detectable signal) is the signal level that is equal to the receiver noise floor in a specified bandwidth.

E4C08 How might lowering the noise figure affect receiver performance?

A. It would reduce the signal to noise ratio
B. It would increase signal to noise ratio
C. It would reduce bandwidth
D. It would increase bandwidth

B   Noise figure is a measure of how much noise is generated in the receiver. If noise figure is reduced, the receiver's noise contribution is reduced. That improves (increases) the signal-to-noise ratio.

E4C09 Which of the following is most likely to be the limiting condition for sensitivity in a modern communications receiver operating at 14 MHz?

A. The noise figure of the RF amplifier
B. Mixer noise
C. Conversion noise
D. Atmospheric noise

D   You are looking for a factor that makes the biggest contribution to received noise. Below 20 MHz, the largest contributor of noise is external to the receiver — atmospheric noise.

E4C10 Which of the following is a desirable amount of selectivity for an amateur RTTY HF receiver?

A. 100 Hz
B. 300 Hz
C. 6000 Hz
D. 2400 Hz

B   You'll want a filter that's wide enough to pass the mark and space frequencies along with their respective sidebands. These are spaced less than 200 Hz apart in the usual RTTY signal. You'll want a little bit extra bandwidth to allow for tuning error. Don't make it too wide or you let through noise and interference.

E4C11 Which of the following is a desirable amount of selectivity for an amateur single-sideband phone receiver?

A. 1 kHz
B. 2.4 kHz
C. 4.2 kHz
D. 4.8 kHz

B Intelligibility of a voice signal is mostly contained in the range of 300 Hz to 2700 Hz. While wider filters may increase fidelity, they're not necessary and they'll allow more noise and interference to pass. Under crowded band conditions, filters with a bandwidth as low as 1.5 kHz can be useful, but at the cost of greatly reduced fidelity.

E4C12 What is an undesirable effect of using too wide a filter bandwidth in the IF section of a receiver?

A. Output-offset overshoot
B. Filter ringing
C. Thermal-noise distortion
D. Undesired signals may be heard

D Some operators like to use wide filters when tuning or monitoring a band that is quiet in terms of noise and in terms of the number of transmitting stations. Few want wider filters when the band is active and many stations are transmitting. A filter that is wider than necessary allows extra signals (and noise) to pass through the IF to the detector and on to the audio output.

E4C13 How does a narrow band roofing filter affect receiver performance?

A. It improves sensitivity by reducing front end noise
B. It improves intelligibility by using low Q circuitry to reduce ringing
C. It improves dynamic range by keeping strong signals near the receive frequency out of the IF stages
D. All of these choice are correct

C The function of a roofing filter is to remove strong nearby signals that may overload the receiver circuits before being rejected by the narrower single-signal filters.

E4C14 Which of these choices is a desirable amount of selectivity for an amateur VHF FM receiver?

A. 1 kHz
B. 2.4 kHz
C. 4.2 kHz
D. 15 kHz

D You'll want a filter with approximately the bandwidth of the FM phone (voice) signal, which is 15 kHz.

E4C15 What is the primary source of noise that can be heard from an HF-band receiver with an antenna connected?

A. Detector noise
B. Induction motor noise
C. Receiver front-end noise
D. Atmospheric noise

D See the discussion for E4C09.

E4D Receiver performance characteristics, part 2: blocking dynamic range, intermodulation and cross-modulation interference; 3rd order intercept; desensitization; preselection

E4D01 What is meant by the blocking dynamic range of a receiver?

A. The difference in dB between the level of an incoming signal which will cause 1 dB of gain compression, and the level of the noise floor
B. The minimum difference in dB between the levels of two FM signals which will cause one signal to block the other
C. The difference in dB between the noise floor and the third order intercept point
D. The minimum difference in dB between two signals which produce third order intermodulation products greater than the noise floor

A Blocking dynamic range (BDR) refers to the ability of a receiver to respond linearly to strong signals. The definition of BDR is the difference between MDS and the input signal level at which receiver gain drops (gain compression) by 1 dB.

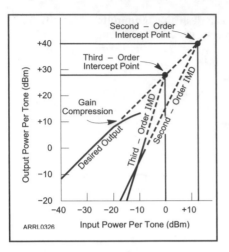

Figure E4D02 — This graph shows how the output power of the receiver for the desired signal and the output power for the second and third-order distortion products vary with changes of the input signal power. The input consists of two equal-power sine-wave signals. Higher intercept points represent better receiver IMD performance.

E4D02 Which of the following describes two types of problems caused by poor dynamic range in a communications receiver?

A. Cross modulation of the desired signal and desensitization from strong adjacent signals
B. Oscillator instability requiring frequent retuning, and loss of ability to recover the opposite sideband, should it be transmitted
C. Cross modulation of the desired signal and insufficient audio power to operate the speaker
D. Oscillator instability and severe audio distortion of all but the strongest received signals

A A receiver with poor IMD dynamic range will exhibit cross modulation of the desired signal by strong adjacent signals. One with poor blocking dynamic range will suffer from desensitization.

E4D03 How can intermodulation interference between two repeaters occur?

A. When the repeaters are in close proximity and the signals cause feedback in one or both transmitter final amplifiers
B. When the repeaters are in close proximity and the signals mix in one or both transmitter final amplifiers
C. When the signals from the transmitters are reflected out of phase from airplanes passing overhead
D. When the signals from the transmitters are reflected in phase from airplanes passing overhead

B Intermodulation can be a problem in transmitters as well as receivers. This can happen when two transmitters are in close proximity and the signals mix in one or both of their final amplifiers or in a non-linear device or junction near the transmitters. This can result in severe interference.

E4D04 What is an effective way to reduce or eliminate intermodulation interference between two repeater transmitters operating in close proximity to one another?

A. By installing a band-pass filter in the feed line between the transmitter and receiver
B. By installing a properly terminated circulator at the output of the transmitter
C. By using a Class C final amplifier
D. By using a Class D final amplifier

B Circulators and isolators are usually highly effective in eliminating intermodulation between two transmitters. They work like one-way valves, allowing energy to flow from the transmitter to the antenna while greatly reducing energy flow in the opposite direction. You might think that installing some type of filter would cure the problem, but that's not true in this case because the offending transmitter will have a very strong signal in the desired passband.

E4D05 If a receiver tuned to 146.70 MHz receives an intermodulation-product signal whenever a nearby transmitter transmits on 146.52 MHz, what are the two most likely frequencies for the other interfering signal?

A. 146.34 MHz and 146.61 MHz
B. 146.88 MHz and 146.34 MHz
C. 146.10 MHz and 147.30 MHz
D. 73.35 MHz and 239.40 MHz

A The frequencies of the strongest IMD components come from the equations

$$f_{IMD} = 2f_1 \pm f_2$$

and

$$f_{IMD} = 2f_2 \pm f_1$$

You are given $f_1 = 146.52$ MHz and $f_{IMD} = 146.70$ MHz, whose sums fall into the UHF range. For that reason you'll only need to look at the differences. Solve for f_2 and you'll find

$$f_2 = 2f_1 - f_{IMD} = (2 \times 146.52 \text{ MHz}) - 146.70 \text{ MHz} = 146.34 \text{ MHz}$$

and

$$f_2 = \frac{f_1 + f_{IMD}}{2} = \frac{146.52 \text{ MHz} + 146.70 \text{ MHz}}{2} = 146.61 \text{ MHz}$$

E4D06   If the signals of two transmitters mix together in one or both of their final amplifiers, and unwanted signals at the sum and difference frequencies of the original signals are generated, what is this called?

A. Amplifier desensitization
B. Neutralization
C. Adjacent channel interference
D. Intermodulation interference

D   Intermodulation interference occurs when the signals of two transmitters mix together in one or both of their final amplifiers and unwanted signals at the sum and difference frequencies of the original signals are generated.

E4D07   Which of the following describes the most significant effect of an off-frequency signal when it is causing cross-modulation interference to a desired signal?

A. A large increase in background noise
B. A reduction in apparent signal strength
C. The desired signal can no longer be heard
D. The off-frequency unwanted signal is heard in addition to the desired signal

D   The term "cross-modulation" is used when modulation from an unwanted signal is heard in addition to the desired signal.

E4D08   What causes intermodulation in an electronic circuit?

A. Too little gain
B. Lack of neutralization
C. Nonlinear circuits or devices
D. Positive feedback

C   In a linear circuit, the output is a faithful representation of the input. Nonlinearities in either circuits or devices cause distortion. This nonlinearity is the cause of intermodulation in an electronic circuit.

E4D09 What is the purpose of the preselector in a communications receiver?

A. To store often-used frequencies
B. To provide a range of AGC time constants
C. To improve rejection of unwanted signals
D. To allow selection of the optimum RF amplifier device

C Communications receivers are often operated in an environment where very strong out-of-band signals are present, such as from commercial or military stations or shortwave broadcast stations. These signals can be so strong that they overload the receiver input circuits, causing desensitization and numerous spurious signals. A preselector is a relatively broad filter that attenuates these out-of-band signals, helping the receiver filter them out without being overloaded.

E4D10 What does a third-order intercept level of 40 dBm mean with respect to receiver performance?

A. Signals less than 40 dBm will not generate audible third-order intermodulation products
B. The receiver can tolerate signals up to 40 dB above the noise floor without producing third-order intermodulation products
C. A pair of 40 dBm signals will theoretically generate the same output on the third order intermodulation frequency as on the input frequency
D. A pair of 1 mW input signals will produce a third-order intermodulation product which is 40 dB stronger than the input signal

C The third-order intercept point of a receiver is that input level where third-order IMD products equal the desired (first-order) output level. Real-world receivers will be in compression or saturation before the input rises to that level, and for that reason you have to compute the intercept point. You can't measure it directly.

You can see how this works in Figure E4D02. The desired output increases 1 dB for a 1-dB increase of the input. The second-order IMD increases 2 dB, and third order goes up 3 dB for each 1 dB increase in input level.

A third-order intercept point of 40 dBm means that it would take a pair of signals with levels of 40 dBm (10 watts!) to create intermodulation products with the same strength as the receiver output on the desired frequency.

E4D11 **Why are third-order intermodulation products within a receiver of particular interest compared to other products?**

A. The third-order product of two signals which are in the band is itself likely to be within the band
B. The third-order intercept is much higher than other orders
C. Third-order products are an indication of poor image rejection
D. Third-order intermodulation produces three products for every input signal

A The frequencies of the strongest IMD components come from the equations

$$f_{IMD} = 2f_1 \pm f_2$$

and

$$f_{IMD} = 2f_2 \pm f_1$$

These frequencies are close to the frequencies of the signals generating the IMD and are likely to be in the same band as the desired signal. So third-order IMD products are the most likely to cause interference.

E4D12 **What is the term for the reduction in receiver sensitivity caused by a strong signal near the received frequency?**

A. Desensitization
B. Quieting
C. Cross-modulation interference
D. Squelch gain rollback

A Desensitization or "desense" is caused by a reduction in gain, or gain compression, that causes the receiver output to drop as if it was less sensitive.

E4D13 **Which of the following can cause receiver desensitization?**

A. Audio gain adjusted too low
B. Strong adjacent-channel signals
C. Audio bias adjusted too high
D. Squelch gain adjusted too low

B Desensitization from a strong signal near the desired signal causes the receiver circuits to be overloaded such that they can no longer amplify the desired signal by the correct amount.

E4D14 Which of the following is a way to reduce the likelihood of receiver desensitization?

A. Decrease the RF bandwidth of the receiver
B. Raise the receiver IF frequency
C. Increase the receiver front end gain
D. Switch from fast AGC to slow AGC

A Removing or attenuating the strong nearby signals is the best way of eliminating receiver desensitization. This is the function of roofing filters that remove strong signals near the desired signal before a single-signal filter determines the receiver's final selectivity.

E4E   Noise suppression: system noise; electrical appliance noise; line noise; locating noise sources; DSP noise reduction; noise blankers

E4E01 Which of the following types of receiver noise can often be reduced by use of a receiver noise blanker?

A. Ignition Noise
B. Broadband "white" noise
C. Heterodyne interference
D. All of these choices are correct

A Noise blankers work best on impulse noise created by sharp, short pulses. Of the noise types listed, ignition noise from the firing of a vehicle's spark plugs is the best candidate for removal by a noise blanker.

E4E02 Which of the following types of receiver noise can often be reduced with a DSP noise filter?

A. Broadband "white" noise
B. Ignition noise
C. Power line noise
D. All of these choices are correct

D DSP (Digital Signal Processing) noise filters can handle a wider range of noise characteristics than analog filters. All three noise types listed here can be removed by DSP filters.

E4E03 Which of the following signals might a receiver noise blanker be able to remove from desired signals?

A. Signals which are constant at all IF levels
B. Signals which appear correlated across a wide bandwidth
C. Signals which appear at one IF but not another
D. Signals which have a sharply peaked frequency distribution

B   Noise blankers work by detecting signals that simultaneously appear across a wide bandwidth, the characteristics of impulse noise pulses.

E4E04 How can conducted and radiated noise caused by an automobile alternator be suppressed?

A. By installing filter capacitors in series with the DC power lead and by installing a blocking capacitor in the field lead
B. By connecting the radio to the battery by the longest possible path and installing a blocking capacitor in both leads
C. By installing a high-pass filter in series with the radio's power lead and a low-pass filter in parallel with the field lead
D. By connecting the radio's power leads directly to the battery and by installing coaxial capacitors in line with the alternator leads

D   Conducted and radiated noise caused by an automobile alternator can be suppressed by connecting the radio's power leads directly to the battery and by installing coaxial capacitors in the alternator leads.

E4E05 How can noise from an electric motor be suppressed?

A. By installing a ferrite bead on the AC line used to power the motor
B. By installing a brute-force AC-line filter in series with the motor leads
C. By installing a bypass capacitor in series with the motor leads
D. By using a ground-fault current interrupter in the circuit used to power the motor

B   A brute-force ac line filter in series with the motor leads can suppress noise from an electric motor.

E4E06 What is a major cause of atmospheric static?

A. Solar radio frequency emissions
B. Thunderstorms
C. Geomagnetic storms
D. Meteor showers

B   Thunderstorms are a major cause of atmospheric static. The other options presented here are not.

E4E07 How can you determine if line-noise interference is being generated within your home?

A. By checking the power-line voltage with a time-domain reflectometer
B. By observing the AC power line waveform with an oscilloscope
C. By turning off the AC power line main circuit breaker and listening on a battery-operated radio
D. By observing the AC power line voltage with a spectrum analyzer

C If the line noise disappears when the ac power to the home is removed, then you should look there for the noise source.

E4E08 What type of signal is picked up by electrical wiring near a radio transmitter?

A. A common-mode signal at the frequency of the radio transmitter
B. An electrical-sparking signal
C. A differential-mode signal at the AC power line frequency
D. Harmonics of the AC power line frequency

A If the transmitter is operating properly, the interfering signal will be at the frequency of the transmitter. "Common mode" means that the electrical wiring is working like an antenna and the signal is being picked up by all of the conductors "in common."

E4E09 What undesirable effect can occur when using an IF type noise blanker?

A. Received audio in the speech range might have an echo effect
B. The audio frequency bandwidth of the received signal might be compressed
C. Nearby signals may appear to be excessively wide even if they meet emission standards
D. FM signals can no longer be demodulated

C Because noise blankers detect strong signals across a wide bandwidth, strong local signals can cause the receiver blanking circuit to activate by mistake. This causes the strong signal to sound distorted in the receiver and also causes distortion of nearby signals, making them appear to have an excessively wide bandwidth. If a strong signal seems to have excessive bandwidth, turn off the receiver noise blanker to see if that might be the problem.

E4E10 What is a common characteristic of interference caused by a "touch controlled" electrical device?

A. The interfering signal sounds like AC hum on an AM receiver or a carrier modulated by 60 Hz FM on a SSB or CW receiver
B. The interfering signal may drift slowly across the HF spectrum
C. The interfering signal can be several kHz in width and usually repeats at regular intervals across a HF band
D. All of these answers are correct

D Touch-controlled devices operate by detecting a shift in an oscillator's frequency when the device is touched. With poor ac line filtering (or none at all), these oscillators often generate interference at harmonics of their operating frequency that may be modulated by the ac power 60 Hz frequency. The oscillators are also rather unstable and so the interference often drifts in frequency.

E4E11 What is the most likely cause if you are hearing combinations of local AM broadcast signals inside one or more of the MF or HF ham bands?

A. The broadcast station is transmitting an over-modulated signal
B. Nearby corroded metal joints are mixing and re-radiating the BC signals
C. You are receiving sky-wave signals from a distant station
D. Your station receiver IF amplifier stage is defective

B Local AM broadcast stations generate very strong signals that can cause current to flow in any metal conductor, such as fences, gutters, metal roofs and siding, pipes, and so forth. Because metal joints exposed to the weather are often corroded and nonlinear, they act as mixers, creating mixing products at various combinations of the broadcast station frequencies.

E4E12 What is one disadvantage of using some automatic DSP notch-filters when attempting to copy CW signals?

A. The DSP filter can remove the desired signal at the same time as it removes interfering signals
B. Any nearby signal passing through the DSP system will always overwhelm the desired signal
C. Received CW signals will appear to be modulated at the DSP clock frequency
D. Ringing in the DSP filter will completely remove the spaces between the CW characters

A A CW signal can appear much like an interfering carrier to a DSP notch filter. The result is that the filter will automatically remove the CW signal along with the properly-removed tone. For that reason, using the auto-notch filters doesn't usually work well on CW.

E4E13 What might be the cause of a loud "roaring" or "buzzing" AC line type of interference that comes and goes at intervals?

A. Arcing contacts in a thermostatically controlled device
B. A defective doorbell or doorbell transformer inside a nearby residence
C. A malfunctioning illuminated advertising display
D. All of these answers are correct

D Power line noise, especially if the source is nearby, can be very loud and contain many harmonics of the 60 Hz line frequency. The resulting signal has the roaring, buzzing quality mentioned in the question. Along with sources such as defective power line insulators, line noise can also be generated by any kind of electrical contacts, switches, or lighting equipment. The arcing associated with this type of noise can be a fire hazard in a business or home.

E4E14 What is one type of electrical interference that might be caused by the operation of a nearby personal computer?

A. A loud AC hum in the audio output of your station receiver
B. A clicking noise at intervals of a few seconds
C. The appearance of unstable modulated or unmodulated signals at specific frequencies
D. A whining type noise that continually pulses off and on

C Interference from a computer (or any microprocessor-controlled device) usually consists of signals created by the sharp rise and fall of the myriad digital signals in the computer. These signals are unmodulated, so they are heard as tones. The signal frequencies change as the digital signal changes and as the computer's internal clock frequencies change with temperature.

Electrical Principles

There will be four questions on your Extra class examination from the Electrical Principles subelement. These four questions will be taken from the four groups of questions labeled E5A through E5D.

E5A Resonance and Q: characteristics of resonant circuits: series and parallel resonance; Q; half-power bandwidth; phase relationships in reactive circuits

E5A01 What can cause the voltage across reactances in series to be larger than the voltage applied to them?

A. Resonance
B. Capacitance
C. Conductance
D. Resistance

A At resonance, the voltage across the inductor and capacitor in a series circuit can be many times greater than the applied voltage. In practical circuits, it can be 10 or 100 times greater. How can this be? The reason is that the capacitor and inductor store the supplied energy, dissipating a small amount of it in resistive losses. The applied voltage "pumps" the resonant circuit, building up energy in each component.

E5A02 What is resonance in an electrical circuit?

A. The highest frequency that will pass current
B. The lowest frequency that will pass current
C. The frequency at which the capacitive reactance equals the inductive reactance
D. The frequency at which the reactive impedance equals the resistive impedance

C At resonance, capacitive and inductive reactances are equal.

E5A03   What is the magnitude of the impedance of a series R-L-C circuit at resonance?

A. High, as compared to the circuit resistance
B. Approximately equal to capacitive reactance
C. Approximately equal to inductive reactance
D. Approximately equal to circuit resistance

D   In a series circuit at resonance, the reactance of L and C cancel, leaving only the circuit resistance.

E5A04   What is the magnitude of the impedance of a circuit with a resistor, an inductor and a capacitor all in parallel, at resonance?

A. Approximately equal to circuit resistance
B. Approximately equal to inductive reactance
C. Low, as compared to the circuit resistance
D. Approximately equal to capacitive reactance

A   In a parallel circuit the basic principle of resonance does not change: the L and C reactances cancel so all that remains is the circuit resistance.

E5A05   What is the magnitude of the current at the input of a series R-L-C circuit as the frequency goes through resonance?

A. Minimum
B. Maximum
C. R/L
D. L/R

B   Current in a series circuit is maximum at resonance because the inductive and capacitive reactances cancel.

E5A06   What is the magnitude of the circulating current within the components of a parallel L-C circuit at resonance?

A. It is at a minimum
B. It is at a maximum
C. It equals 1 divided by the quantity [2 multiplied by Pi, multiplied by the square root of (inductance "L" multiplied by capacitance "C")]
D. It equals 2 multiplied by Pi, multiplied by frequency "F", multiplied by inductance "L"

B   The circulating current within the components of a parallel L-C circuit are maximum at resonance. Circulating current transfers the circuit's stored energy back and forth between the inductive and capacitive reactance. When the two types of reactance are equal or balanced at resonance, the circulating current is at a maximum.

E5A07   What is the magnitude of the current at the input of a parallel R-L-C circuit at resonance?

A. Minimum
B. Maximum
C. R/L
D. L/R

A   At resonance, the circulating current of a parallel circuit is out of phase with the applied current. The effect is that no current flows through the resonant circuit and its impedance is very high. Current in a parallel R-L-C circuit does not go to zero at resonance because of the remaining resistance.

E5A08   What is the phase relationship between the current through and the voltage across a series resonant circuit?

A. The voltage leads the current by 90 degrees
B. The current leads the voltage by 90 degrees
C. The voltage and current are in phase
D. The voltage and current are 180 degrees out of phase

C   At resonance, the impedance of the circuit is completely resistive because the inductive and capacitive reactances have cancelled. In a resistive circuit, voltage and current are always in phase.

E5A09   What is the phase relationship between the current through and the voltage across a parallel resonant circuit?

A. The voltage leads the current by 90 degrees
B. The current leads the voltage by 90 degrees
C. The voltage and current are in phase
D. The voltage and current are 180 degrees out of phase

C   See E5A08. It doesn't matter whether the circuit is a series or parallel resonant circuit.

E5A10 **What is the half-power bandwidth of a parallel resonant circuit that has a resonant frequency of 1.8 MHz and a Q of 95?**

A. 18.9 kHz
B. 1.89 kHz
C. 94.5 kHz
D. 9.45 kHz

A This is the first of several questions dealing with the bandwidth of resonant circuits. The relationship between bandwidth, BW, resonant frequency, f_0, and quality factor, Q, is:

$$BW = \frac{f_0}{Q}$$

Using the values in the question you'll find

$$BW = \frac{1.8 \times 10^6 \text{ Hz}}{95} = 18.9 \text{ kHz}$$

E5A11 **What is the half-power bandwidth of a parallel resonant circuit that has resonant frequency of 7.1 MHz and a Q of 150?**

A. 157.8 Hz
B. 315.6 Hz
C. 47.3 kHz
D. 23.67 kHz

C Use the same relationship given the previous question to solve for the bandwidth. This time instead of using Hz, use kHz for the resonant frequency. That will give you the answer directly in kHz. Using the values given, compute the bandwidth

$$BW = \frac{f_0}{Q} = \frac{7100 \text{ kHz}}{150} = 47.3 \text{ kHz}$$

E5A12 **What is the half-power bandwidth of a parallel resonant circuit that has a resonant frequency of 3.7 MHz and a Q of 118?**

A. 436.6 kHz
B. 218.3 kHz
C. 31.4 kHz
D. 15.7 kHz

C Try solving this without performing the exact calculation. If you divide the frequency by the Q, you get approximately 30 kHz. The closest answer is C, 31.4 kHz.

E5A13 What is the half-power bandwidth of a parallel resonant circuit that has a resonant frequency of 14.25 MHz and a Q of 187?

A. 38.1 kHz
B. 76.2 kHz
C. 1.332 kHz
D. 2.665 kHz

B Using the same formula, B is seen to be the answer.

E5A14 What is the resonant frequency of a series RLC circuit if R is 22 ohms, L is 50 microhenrys and C is 40 picofarads?

A. 44.72 MHz
B. 22.36 MHz
C. 3.56 MHz
D. 1.78 MHz

C The resonant frequency, f_0, is the frequency at which X_L and X_C are equal. That can be shown to be

$$f_0 = \frac{1}{2\pi\sqrt{LC}}$$

The equation is the same for parallel and series resonant circuits because the only requirement is that the two reactances are equal, regardless of how they are connected together. Converting μH to H and pF to F,

$$f_0 = \frac{1}{6.28 \times \sqrt{50 \times 10^{-6} \times 40 \times 10^{-12}}} = 3{,}560{,}617 \text{ Hz} = 3.56 \text{ MHz}$$

Note that the value of R does not affect the resonant frequency of the circuit.

E5A15 What is the resonant frequency of a series RLC circuit if R is 56 ohms, L is 40 microhenrys and C is 200 picofarads?

A. 3.76 MHz
B. 1.78 MHz
C. 11.18 MHz
D. 22.36 MHz

B Using the same formula:

$$f_0 = \frac{1}{6.28 \times \sqrt{40 \times 10^{-6} \times 200 \times 10^{-12}}} = 1.78 \text{ MHz}$$

E5A16 What is the resonant frequency of a parallel RLC circuit if R is 33 ohms, L is 50 microhenrys and C is 10 picofarads?

    A. 23.5 MHz
    B. 23.5 kHz
    C. 7.12 kHz
    D. 7.12 MHz

D   Using the same formula:

$$f_0 = \frac{1}{6.28 \times \sqrt{50 \times 10^{-6} \times 10 \times 10^{-12}}} = 7.12\text{ MHz}$$

Watch out because C and D both have the same numeric answer, but C shows kHz! This is an easy one to miss if you're not careful.

E5A17 What is the resonant frequency of a parallel RLC circuit if R is 47 ohms, L is 25 microhenrys and C is 10 picofarads?

    A. 10.1 MHz
    B. 63.2 MHz
    C. 10.1 kHz
    D. 63.2 kHz

A   Using the same formula:

$$f_0 = \frac{1}{6.28 \times \sqrt{25 \times 10^{-6} \times 10 \times 10^{-12}}} = 10.1\text{ MHz}$$

Again, pay attention to the units of kHz or MHz.

E5B Time constants and phase relationships: R/L/C time constants: definition; time constants in RL and RC circuits; phase angle between voltage and current; phase angles of series and parallel circuits

E5B01 What is the term for the time required for the capacitor in an RC circuit to be charged to 63.2% of the supply voltage?

A. An exponential rate of one
B. One time constant
C. One exponential period
D. A time factor of one

B In an RC circuit, when the capacitor has no initial charge, it takes one time constant to charge the capacitor to 63.2% of the applied voltage. In the graph, you can see how the voltage across a capacitor rises with time, when charged through a resistor. The symbol τ is used to indicate a period equal to the time constant.

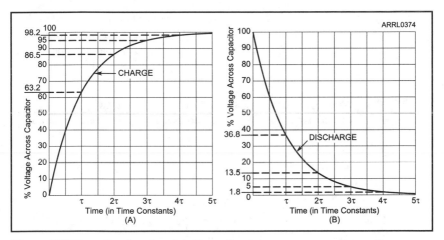

Figure E5B01 — Graphs showing the voltage across a capacitor as it charges (A) and discharges (B) through a resistor. The time it takes the capacitor to charge to 63.2% of the applied voltage or to discharge to 36.8% of an initial voltage is the circuit's time constant. The symbol τ is used for time constant.

E5B02 What is the term for the time it takes for a charged capacitor in an RC circuit to discharge to 36.8% of its initial value of stored charge?

A. One discharge period
B. An exponential discharge rate of one
C. A discharge factor of one
D. One time constant

D The voltage while charging and discharging has the same general exponential shape, so it also takes one time constant to discharge 63.2% of the initial voltage, which leaves 36.8% of the initial voltage across the capacitor.

E5B03 The capacitor in an RC circuit is discharged to what percentage of the starting voltage after two time constants?

A. 86.5%
B. 63.2%
C. 36.8%
D. 13.5%

D The voltage after discharging for one time constant can be treated as the initial voltage for the second time constant. The voltage remaining after two time constants, then, is 36.8% of 36.8% = 13.5%. The discharge graph shows the result.

E5B04 What is the time constant of a circuit having two 220-microfarad capacitors and two 1-megohm resistors all in parallel?

A. 55 seconds
B. 110 seconds
C. 440 seconds
D. 220 seconds

D In this parallel circuit, the total resistance is 500 kΩ. The total capacitance is 440 μF. The time constant is

$\tau = R\,C = (500 \times 10^3) \times (440 \times 10^{-6}) = 220$ seconds

E5B05 How long does it take for an initial charge of 20 V DC to decrease to 7.36 V DC in a 0.01-microfarad capacitor when a 2-megohm resistor is connected across it?

A. 0.02 seconds
B. 0.04 seconds
C. 20 seconds
D. 40 seconds

A First you'll need to find out how many time constants this amount of discharge represents. Because 7.36 V is 36.8% of 20 V (7.36/20), the circuit has discharged for one time constant as you can see on the discharge curve. The time constant for this circuit is

$$\tau = R\,C = (2 \times 10^6) \times (0.01 \times 10^{-6}) = 0.02 \text{ second}$$

E5B06 How long does it take for an initial charge of 800 V DC to decrease to 294 V DC in a 450-microfarad capacitor when a 1-megohm resistor is connected across it?

A. 4.50 seconds
B. 9 seconds
C. 450 seconds
D. 900 seconds

C As in the previous question, find out how many time constants the discharge represents. 294 V is 36.8% of 800 V so the circuit has discharged by one time constant. The time constant for the circuit is

$$\tau = R\,C = (1 \times 10^6) \times (450 \times 10^{-6}) = 450 \text{ seconds}$$

E5B07 **What is the phase angle between the voltage across and the current through a series R-L-C circuit if XC is 500 ohms, R is 1 kilohm, and XL is 250 ohms?**

A. 68.2 degrees with the voltage leading the current
B. 14.0 degrees with the voltage leading the current
C. 14.0 degrees with the voltage lagging the current
D. 68.2 degrees with the voltage lagging the current

C (Before beginning this series of questions, you should be familiar with polar coordinates in subelement E5C below.) The total reactance in this series configuration is $250 \, \Omega - 500 \, \Omega = -250 \, \Omega$. The phase angle between the voltage and the current is

$$\tan^{-1}\left(\frac{X}{R}\right) = \tan^{-1}\left(\frac{-250 \, \Omega}{1000 \, \Omega}\right) = -14.0°$$

Because the angle is negative, the voltage lags the current. Since the net reactance is negative, the phase angle needs to be negative. Because the net reactance is smaller than the resistance, the phase angle will be less than 45°.

E5B08 **What is the phase angle between the voltage across and the current through a series R-L-C circuit if XC is 100 ohms, R is 100 ohms, and XL is 75 ohms?**

A. 14 degrees with the voltage lagging the current
B. 14 degrees with the voltage leading the current
C. 76 degrees with the voltage leading the current
D. 76 degrees with the voltage lagging the current

A The total reactance in this series configuration is $75 \, \Omega - 100 \, \Omega$. The phase angle between the voltage and the current is

$$\tan^{-1}\left(\frac{-25 \, \Omega}{100 \, \Omega}\right) = -14°$$

Again, voltage lags current because the phase angle is negative.

E5B09   What is the relationship between the current through and the voltage across a capacitor?

A. Voltage and current are in phase
B. Voltage and current are 180 degrees out of phase
C. Voltage leads current by 90 degrees
D. Current leads voltage by 90 degrees

D   The current leads voltage by 90°. Have you heard of "Eli the iceman"? It's an easy way to remember current and voltage relationships in reactive circuits. ELI means voltage (E) leads current (I) in an inductance (L). This case is ICE — current (I) leads voltage (E) in a capacitor (C). For a pure reactance (no resistance), the phase angle is 90°.

E5B10   What is the relationship between the current through an inductor and the voltage across an inductor?

A. Voltage leads current by 90 degrees
B. Current leads voltage by 90 degrees
C. Voltage and current are 180 degrees out of phase
D. Voltage and current are in phase

A   This is the complement to the previous question. In other words, this is "ELI." That means voltage leads current. Because this is pure reactance, the phase shift is 90°.

E5B11   What is the phase angle between the voltage across and the current through a series RLC circuit if XC is 25 ohms, R is 100 ohms, and XL is 50 ohms?

A. 14 degrees with the voltage lagging the current
B. 14 degrees with the voltage leading the current
C. 76 degrees with the voltage lagging the current
D. 76 degrees with the voltage leading the current

B   The total reactance in this series configuration is 50 Ω – 25 Ω. The phase angle between the voltage and the current is

$$\tan^{-1}\left(\frac{25\,\Omega}{100\,\Omega}\right) = 14°$$

The positive angle means that voltage leads current.

To review our rules of thumb: Since the net reactance is positive, the phase angle needs to be positive; because the net reactance is smaller than the resistance, the phase angle needs to be less than 45°.

E5B12 What is the phase angle between the voltage across and the current through a series RLC circuit if XC is 75 ohms, R is 100 ohms, and XL is 50 ohms?

A. 76 degrees with the voltage lagging the current
B. 14 degrees with the voltage leading the current
C. 14 degrees with the voltage lagging the current
D. 76 degrees with the voltage leading the current

C The total reactance in this series configuration is 50 Ω – 75 Ω. The phase angle between the voltage and the current is

$$\tan^{-1}\left(\frac{-25\ \Omega}{100\ \Omega}\right) = -14°$$

A negative phase angle means that voltage lags current.

E5B13 What is the phase angle between the voltage across and the current through a series RLC circuit if XC is 250 ohms, R is 1 kilohm, and XL is 500 ohms?

A. 81.47 degrees with the voltage lagging the current
B. 81.47 degrees with the voltage leading the current
C. 14.04 degrees with the voltage lagging the current
D. 14.04 degrees with the voltage leading the current

D The total reactance in this series configuration is 500 Ω – 250 Ω. The phase angle between the voltage and the current is

$$\tan^{-1}\left(\frac{250\ \Omega}{1000\ \Omega}\right) = 14.04°$$

A positive phase angle means that voltage is leading current.

E5C Impedance plots and coordinate systems: plotting impedances in polar coordinates; rectangular coordinates

E5C01 In polar coordinates, what is the impedance of a network consisting of a 100-ohm-reactance inductor in series with a 100-ohm resistor?

A. 121 ohms at an angle of 35 degrees
B. 141 ohms at an angle of 45 degrees
C. 161 ohms at an angle of 55 degrees
D. 181 ohms at an angle of 65 degrees

B In this section, you have a series of questions dealing with series resistance and rectangular-to-polar coordinate conversion. In rectangular coordinates, impedance is the sum of the resistive and reactive components. You plot impedance with resistance on the X axis and reactance on the Y axis.

In polar coordinates, impedance is described by magnitude and the corresponding phase angle. You can calculate these from resistance (R) and reactance (X) using the following formulas, in which the vertical bars around Z indicate "magnitude" without regard to the phase angle.

$$|Z| = \sqrt{R^2 + X^2}$$

and

$$\theta = \tan^{-1}\left(\frac{X}{R}\right)$$

Solving these equations with the values in the question, you find

$$|Z| = \sqrt{100^2 + 100^2} = 141\ \Omega$$

and

$$\theta = \tan^{-1}\left(\frac{100}{100}\right) = 45°$$

E5C02 In polar coordinates, what is the impedance of a network consisting of a 100-ohm-reactance inductor, a 100-ohm-reactance capacitor, and a 100-ohm resistor, all connected in series?

A. 100 ohms at an angle of 90 degrees
B. 10 ohms at an angle of 0 degrees
C. 10 ohms at an angle of 90 degrees
D. 100 ohms at an angle of 0 degrees

D   Here's a chance to review resonance that you studied earlier. This is a resonant circuit because the capacitive and inductive reactances are equal. That means that the reactances cancel and you have only the resistance left, which is an impedance of 100 Ω with no reactive phase shift (angle = 0°).

E5C03 In polar coordinates, what is the impedance of a network consisting of a 300-ohm-reactance capacitor, a 600-ohm-reactance inductor, and a 400-ohm resistor, all connected in series?

A. 500 ohms at an angle of 37 degrees
B. 900 ohms at an angle of 53 degrees
C. 400 ohms at an angle of 0 degrees
D. 1300 ohms at an angle of 180 degrees

A   Remember that capacitive reactance is negative. Use the formulas for converting to polar coordinates and find

$$|Z| = \sqrt{400^2 + (600 - 300)^2} = 500\ \Omega$$

and

$$\theta = \tan^{-1}\left(\frac{600 - 300}{400}\right) = 37°$$

E5C04 In polar coordinates, what is the impedance of a network
consisting of a 400-ohm-reactance capacitor in series with a
300-ohm resistor?

A. 240 ohms at an angle of 36.9 degrees
B. 240 ohms at an angle of –36.9 degrees
C. 500 ohms at an angle of 53.1 degrees
D. 500 ohms at an angle of –53.1 degrees

D Use the same two formulas to solve this problem. Remember that
capacitive reactance is negative.

$$|Z| = \sqrt{300^2 + (-400)^2} = 500 \ \Omega$$

and

$$\theta = \tan^{-1}\left(\frac{-400}{300}\right) = -53.1°$$

E5C05 In polar coordinates, what is the impedance of a network consisting of a 400-ohm-reactance inductor in parallel with a 300-ohm resistor?

A. 240 ohms at an angle of 36.9 degrees
B. 240 ohms at an angle of −36.9 degrees
C. 500 ohms at an angle of 53.1 degrees
D. 500 ohms at an angle of −53.1 degrees

A   In this section, you also have questions dealing with parallel resistance and reactance. In parallel circuits, the formulas are

$$|Z| = \frac{1}{\dfrac{1}{R} + \dfrac{1}{X}} = \frac{R \times X}{\sqrt{R^2 + X^2}}$$

and

$$\theta = \tan^{-1}\left(\frac{R}{X}\right)$$

While the formula for the phase angle may look familiar, it is different. Look again and you'll see that R and X have swapped positions when compared to the formula for series circuits. Now, substitute the values into the formulas and you get

$$|Z| = \frac{400 \times 300}{\sqrt{400^2 + 300^2}} = 240\ \Omega$$

and the phase angle is

$$\theta = \tan^{-1}\left(\frac{300}{400}\right) = 36.9°$$

E5C06   In polar coordinates, what is the impedance of a network consisting of a 100-ohm-reactance capacitor in series with a 100-ohm resistor?

A.   121 ohms at an angle of –25 degrees
B.   191 ohms at an angle of –85 degrees
C.   161 ohms at an angle of –65 degrees
D.   141 ohms at an angle of –45 degrees

D   This is another series circuit, so go back to the series formulas to calculate the impedance

$$|Z| = \sqrt{100^2 + (-100)^2} = 141\,\Omega$$

and

$$\theta = \tan^{-1}\left(\frac{-100}{100}\right) = -45°$$

Remember that for the series case, the phase angle is the arctangent of the reactance, X, divided by the resistance, R.

E5C07 In polar coordinates, what is the impedance of a network comprised of a 100-ohm-reactance capacitor in parallel with a 100-ohm resistor?

A. 31 ohms at an angle of -15 degrees
B. 51 ohms at an angle of -25 degrees
C. 71 ohms at an angle of -45 degrees
D. 91 ohms at an angle of -65 degrees

C Let's do another parallel network. The question looks a lot like the previous one, but that was for a series network. This time, you'll use the formulas for parallel circuits (see E5C05), which means the total impedance is

$$|Z| = \frac{100 \times (-100)}{\sqrt{100^2 + (-100)^2}} = 71\,\Omega$$

and the phase angle is

$$\theta = \tan^{-1}\left(\frac{100}{-100}\right) = -45°$$

E5C08 In polar coordinates, what is the impedance of a network comprised of a 300-ohm-reactance inductor in series with a 400-ohm resistor?

A. 400 ohms at an angle of 27 degrees
B. 500 ohms at an angle of 37 degrees
C. 500 ohms at an angle of 47 degrees
D. 700 ohms at an angle of 57 degrees

B Are you ready for one more series circuit? Use the formulas for series and compute the total impedance, which is

$$|Z| = \sqrt{400^2 + 300^2} = 500\,\Omega$$

and

$$\theta = \tan^{-1}\left(\frac{300}{400}\right) = 37°$$

E5C09   When using rectangular coordinates to graph the impedance of a circuit, what does the horizontal axis represent?

A. The voltage or current associated with the resistive component
B. The voltage or current associated with the reactive component
C. The sum of the reactive and resistive components
D. The difference between the resistive and reactive components

A   The horizontal axis represents the resistive component.

E5C10   When using rectangular coordinates to graph the impedance of a circuit, what does the vertical axis represent?

A. The voltage or current associated with the resistive component
B. The voltage or current associated with the reactive component
C. The sum of the reactive and resistive components
D. The difference between the resistive and reactive components

B   The vertical axis represents the reactive component.

E5C11   What do the two numbers represent that are used to define a point on a graph using rectangular coordinates?

A. The magnitude and phase of the point
B. The sine and cosine values
C. The coordinate values along the horizontal and vertical axes
D. The tangent and cotangent values

C   The numbers given to graph a point in rectangular coordinates represent values along the horizontal and vertical axes.

E5C12   If you plot the impedance of a circuit using the rectangular coordinate system and find the impedance point falls on the right side of the graph on the horizontal line, what do you know about the circuit?

A. It has to be a direct current circuit
B. It contains resistance and capacitive reactance
C. It contains resistance and inductive reactance
D. It is equivalent to a pure resistance

D   If the point is on the horizontal axis (resistance), it has no reactive (Y-axis) component. That means you have a pure resistance.

E5C13 What coordinate system is often used to display the resistive, inductive, and/or capacitive reactance components of an impedance?

A. Maidenhead grid
B. Faraday grid
C. Elliptical coordinates
D. Rectangular coordinates

D In the rectangular coordinate system, resistance is plotted on the horizontal axis and reactance on the vertical axis.

E5C14 What coordinate system is often used to display the phase angle of a circuit containing resistance, inductive and/or capacitive reactance?

A. Maidenhead grid
B. Faraday grid
C. Elliptical coordinates
D. Polar coordinates

D This may appear similar to the previous question. However, it is the polar coordinate system that plots phase angle directly.

E5C15 In polar coordinates, what is the impedance of a circuit of 100 − j100 ohms impedance?

A. 141 ohms at an angle of −45 degrees
B. 100 ohms at an angle of 45 degrees
C. 100 ohms at an angle of −45 degrees
D. 141 ohms at an angle of 45 degrees

A Once again, you'll need to convert this to polar coordinates. Do you remember the formulas that you used for series networks? You'll use them again here to calculate impedance

$$|Z| = \sqrt{100^2 + (-100)^2} = 141 \, \Omega$$

and the phase angle is

$$\theta = \tan^{-1}\left(\frac{-100}{100}\right) = -45°$$

E5C16   In polar coordinates, what is the impedance of a circuit that has an admittance of 7.09 millisiemens at 45 degrees?

A. 5.03 x 10 –E05 ohms at an angle of 45 degrees
B. 141 ohms at an angle of –45 degrees
C. 19,900 ohms at an angle of –45 degrees
D. 141 ohms at an angle of 45 degrees

B   Impedance and admittance are reciprocal quantities. When taking the reciprocal in polar coordinates

$$\frac{1}{\text{magnitude } \angle \theta} = \frac{1}{\text{magnitude}} \angle -\theta$$

so

$$\frac{1}{7.09 \text{ mS } \angle 45°} = \frac{1}{0.00709} \angle -45° = 141 \angle -45°$$

E5C17   In rectangular coordinates, what is the impedance of a circuit that has an admittance of 5 millisiemens at –30 degrees?

A. $173 - j100$ ohms
B. $200 + j100$ ohms
C. $173 + j100$ ohms
D. $200 - j100$ ohms

C   Now you'll have to convert polar coordinates to rectangular coordinates. However, first you'll want to convert admittance to impedance, as shown in the answer to the previous question. That means you'll start by taking the reciprocal of the magnitude, which is 1/0.005 S = 200 Ω and the phase angle becomes –(–30) degrees = 30°.

To convert from polar coordinates use

$$R = \cos(\theta) \times Z = \cos(30°) \times 200 = 173 \ \Omega$$

and

$$X = \sin(\theta) \times Z = \sin(30°) \times 200 = +j100 \ \Omega$$

E5C18 In polar coordinates, what is the impedance of a series circuit consisting of a resistance of 4 ohms, an inductive reactance of 4 ohms, and a capacitive reactance of 1 ohm?

A. 6.4 ohms at an angle of 53 degrees
B. 5 ohms at an angle of 37 degrees
C. 5 ohms at an angle of 45 degrees
D. 10 ohms at an angle of –51 degrees

B In this circuit, the total impedance is

$$Z = R + (jX_L - jX_C) = 4\,\Omega + (j4\,\Omega - j1\,\Omega) = 4\,\Omega + j3\,\Omega$$

Use the conversion formulas

$$|Z| = \sqrt{4^2 + 3^2} = 5\,\Omega$$

and the phase angle is

$$\theta = \tan^{-1}\left(\frac{3}{4}\right) = 37°$$

E5C19 Which point on Figure E5-2 best represents that impedance of a series circuit consisting of a 400 ohm resistor and a 38 picofarad capacitor at 14 MHz?

A. Point 2
B. Point 4
C. Point 5
D. Point 6

B In this circuit, the total impedance is

$$Z = R + X = R + \frac{1}{j\,2\pi\,f\,C}$$

Using the values in the question and remembering that $1/j = -j$,

$$Z = 400 + \frac{1}{j6.28 \times (14 \times 10^6) \times (38 \times 10^{-12})} = 400 - j300\,\Omega$$

To find this point, find +400 on the horizontal axis and then move vertically down to the point that corresponds to –300 (remember, capacitive reactance is negative). In the Figure, that corresponds to Point 4.

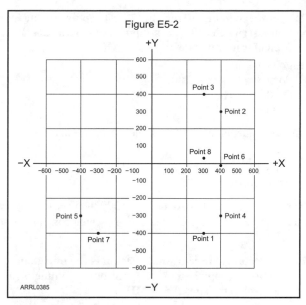

Figure E5-2

Figure E5-2 — This figure is used for questions E5C19, E5C20, E5C21 and E5C23.

E5C20 Which point in Figure E5-2 best represents the impedance of a series circuit consisting of a 300 ohm resistor and an 18 microhenry inductor at 3.505 MHz?

A. Point 1
B. Point 3
C. Point 7
D. Point 8

B In this circuit, the total impedance is

$$Z = R + X = R + j\,2\pi\,f$$

Using the numbers in the question,

$$Z = 300 + j6.28 \times (3.505 \times 10^6) \times (18 \times 10^{-6}) = 300 + j400\ \Omega$$

To find this point, find +300 on the horizontal axis, and then move up (positive direction) to +400 on the vertical axis. This corresponds to Point 3.

E5C21   Which point on Figure E5-2 best represents the impedance of a series circuit consisting of a 300 ohm resistor and a 19 picofarad capacitor at 21.200 MHz?

A. Point 1
B. Point 3
C. Point 7
D. Point 8

A   Using the equation from E5C19, the total impedance of this circuit is

$$Z = 300 + \frac{1}{j\,6.28 \times (21.2 \times 10^6) \times (19 \times 10^{-12})} = 300 - j400\ \Omega$$

Find the point by moving horizontally to +300 then vertically (down) to –400 (remember, capacitive reactance is negative), which corresponds to Point 1.

E5C22   In rectangular coordinates, what is the impedance of a network comprised of a 10-microhenry inductor in series with a 40-ohm resistor at 500 MHz?

A. 40 + j31,400
B. 40 – j31,400
C. 31,400 + j40
D. 31,400 – j40

A   Here's a shortcut. You can immediately eliminate C and D because they have the resistive component where the reactive component belongs. Since this is an inductive circuit, the reactive part has a plus sign, leaving A as the correct answer.

You can verify intuition by analysis: for this circuit, the total impedance is the complex sum of the resistive and inductive portions at the operating frequency or

$$Z = 40 + j\,6.28 \times (500 \times 10^6) \times (10 \times 10^{-6}) = 40 + j31416\ \Omega$$

E5C23 Which point on Figure E5-2 best represents the impedance of a series circuit consisting of a 300-ohm resistor, a 0.64-microhenry inductor and an 85-picofarad capacitor at 24.900 MHz?

    A. Point 1
    B. Point 3
    C. Point 5
    D. Point 8

D The total impedance of this circuit is

$$Z = R + j(X_L - X_C)$$

$$Z = 300 + j\left(2\pi f L - \frac{1}{2\pi f C}\right)$$

$$Z = 300 +$$

$$j\left[[6.28\times(24.9\times10^6)\times(0.64\times10^{-6})] - \frac{1}{6.28\times(24.9\times10^6)\times(85\times10^{-12})}\right]$$
$$Z = 300 + j(100 - 75) = 300 + j\,25\;\Omega$$

Find the point by moving horizontally to +300 then vertically (up) to 25, which corresponds to Point 8.

E5D   AC and RF energy in real circuits: skin effect; electrostatic and electromagnetic fields; reactive power; power factor; coordinate systems

E5D01 What is the result of skin effect?

    A. As frequency increases, RF current flows in a thinner layer of the conductor, closer to the surface
    B. As frequency decreases, RF current flows in a thinner layer of the conductor, closer to the surface
    C. Thermal effects on the surface of the conductor increase the impedance
    D. Thermal effects on the surface of the conductor decrease the impedance

A Skin effect occurs at RF. As the frequency increases, electric and magnetic fields of the signal don't penetrate as deeply into a conductor. This results in RF current flowing in a progressively thinner layer at the surface of the conductor as the frequency increases.

E5D02 Why is the resistance of a conductor different for RF currents than for direct currents?

A. Because the insulation conducts current at high frequencies
B. Because of the Heisenburg Effect
C. Because of skin effect
D. Because conductors are non-linear devices

C At RF current flows mostly along the surface, while direct currents can use the full cross section of a conductor. In other words, the conductor looks larger at dc than it does at RF.

E5D03 What device is used to store electrical energy in an electrostatic field?

A. A battery
B. A transformer
C. A capacitor
D. An inductor

C You can store energy in a capacitor by applying a dc voltage across the terminals. Current will flow onto the capacitor electrodes or plates, creating a voltage between them. This voltage creates an electrostatic field in which electrical energy is stored.

E5D04 What unit measures electrical energy stored in an electrostatic field?

A. Coulomb
B. Joule
C. Watt
D. Volt

B Energy of all types is measured in Joules, which is a measure of capacity to do work.

E5D05 What is a magnetic field?

A. Electric current flow through the space around a permanent magnet
B. The region surrounding a magnet through which a magnetic force acts
C. The space between the plates of a charged capacitor, through which a magnetic force acts
D. The force that drives current through a resistor

B A magnetic field is the space through which a magnetic force acts.

E5D06   In what direction is the magnetic field oriented about a conductor in relation to the direction of electron flow?

A. In the same direction as the current
B. In a direction opposite to the current
C. In all directions; omnidirectional
D. In a direction determined by the left-hand rule

D   A magnetic field curls around electrical current. The direction of a magnetic field around a conductor can be determined by using the left-hand rule. With the wire going across the palm of your left hand, and with your thumb pointed in the direction of electron flow, wrap your fingers around the wire. Your fingers will be pointing in the direction of the magnetic field.

E5D07   What determines the strength of a magnetic field around a conductor?

A. The resistance divided by the current
B. The ratio of the current to the resistance
C. The diameter of the conductor
D. The amount of current

D   Magnetic field strength is proportional to the current, and is stronger when the current is greater.

E5D08   What is the term for energy that is stored in an electromagnetic or electrostatic field?

A. Amperes-joules
B. Potential energy
C. Joules-coulombs
D. Kinetic energy

B   Don't be confused — stored energy is always potential energy as long as it is stored.

E5D09   What is the term for an out-of-phase, nonproductive power associated with inductors and capacitors?

A. Effective power
B. True power
C. Peak envelope power
D. Reactive power

D   The correct term for the out-of-phase, nonproductive power associated with inductors and capacitors is reactive power.

E5D10   In a circuit that has both inductors and capacitors, what happens to reactive power?

A. It is dissipated as heat in the circuit
B. It is repeatedly exchanged between the associated magnetic and electric fields, but is not dissipated
C. It is dissipated as kinetic energy in the circuit
D. It is dissipated in the formation of inductive and capacitive fields

B   The reactive power moves back and forth between the magnetic and electric fields, but is only stored and not dissipated. Only the resistive part of a circuit will dissipate power.

E5D11   How can the true power be determined in an AC circuit where the voltage and current are out of phase?

A. By multiplying the apparent power times the power factor
B. By dividing the reactive power by the power factor
C. By dividing the apparent power by the power factor
D. By multiplying the reactive power times the power factor

A   Power factor, PF, is a quantity that relates the apparent power in a circuit to the real power. You can find the real or true power by multiplying the apparent power by the power factor.

E5D12   What is the power factor of an R-L circuit having a 60 degree phase angle between the voltage and the current?

A. 1.414
B. 0.866
C. 0.5
D. 1.73

C   The power factor is also the cosine of the phase angle between the voltage and the current, and the cosine of 60° is 0.5. The cosine of 0° is 1. When the voltage and current are in phase the power factor is 1 and there is no reactive power — it is all real power. When the voltage and current are out of phase by 90°, the power is all reactive and the power factor is 0 (the cosine of 90°).

E5D13 How many watts are consumed in a circuit having a power factor of 0.2 if the input is 100-V AC at 4 amperes?

A. 400 watts
B. 80 watts
C. 2000 watts
D. 50 watts

B Here you apply the power factor to compute the power from

$P = E \times I \times PF = 100 \text{ V} \times 4A \times 0.2 = 80 \text{ W}$

E5D14 How much power is consumed in a circuit consisting of a 100 ohm resistor in series with a 100 ohm inductive reactance drawing 1 ampere?

A. 70.7 Watts
B. 100 Watts
C. 141.4 Watts
D. 200 Watts

B Start by finding the impedance and phase angle

$$|Z| = \sqrt{100^2 + 100^2} = 141 \ \Omega$$

and

$$\theta = \tan^{-1}\left(\frac{100}{100}\right) = 45°$$

Power factor (PF) = $\cos \theta = \cos 45° = 0.707$

Apparent power = $I^2Z = 1^2 \times 141 = 141$ watts

To find real power, multiply by PF

$P = 141 \times 0.707 = 100$ watts

E5D15 What is reactive power?

A. Wattless, nonproductive power
B. Power consumed in wire resistance in an inductor
C. Power lost because of capacitor leakage
D. Power consumed in circuit Q

A See the discussion for E5D09.

E5D16 What is the power factor of an RL circuit having a 45 degree phase angle between the voltage and the current?

A. 0.866
B. 1.0
C. 0.5
D. 0.707

D Use the formula

Power factor (PF) = cos 45° = 0.707

E5D17 What is the power factor of an RL circuit having a 30 degree phase angle between the voltage and the current?

A. 1.73
B. 0.5
C. 0.866
D. 0.577

C Use the formula

Power factor (PF) = cos 30° = 0.866

E5D18 How many watts are consumed in a circuit having a power factor of 0.6 if the input is 200V AC at 5 amperes?

A. 200 watts
B. 1000 watts
C. 1600 watts
D. 600 watts

D Use the formula

Real power = E × I × PF = 200 × 5 × 0.6 = 600 watts

E5D19 How many watts are consumed in a circuit having a power factor of 0.71 if the apparent power is 500 watts?

A. 704 W
B. 355 W
C. 252 W
D. 1.42 mW

B Use the formula

Real power = Apparent Power × PF = 500 × 0.71 = 355 watts

Circuit Components

There will be six questions on your Extra class examination from the Circuit Components subelement. These six questions will be taken from the six groups of questions labeled E6A through E6F.

E6A Semiconductor materials and devices: semiconductor materials (germanium, silicon, P-type, N-type); transistor types: NPN, PNP, junction, power; field-effect transistors: enhancement mode; depletion mode; MOS; CMOS; N-channel; P-channel

E6A01 In what application is gallium arsenide used as a semiconductor material in preference to germanium or silicon?

A. In high-current rectifier circuits
B. In high-power audio circuits
C. At microwave frequencies
D. At very low frequency RF circuits

C Gallium arsenide (GaAs) has performance advantages for use at microwave frequencies. For that reason, it is often used to make solid-state devices for operation on those frequencies.

E6A02 What type of semiconductor material contains more free electrons than pure germanium or silicon crystals?

A. N-type
B. P-type
C. Bipolar
D. Insulated gate

A Since electrons carry a negative charge, symbolized by N, a semiconductor material that contains more free electrons than pure germanium or silicon crystals electrons is called N-type material.

E6A03 What are the majority charge carriers in P-type semiconductor material?

A. Free neutrons
B. Free protons
C. Holes
D. Free electrons

C P-type material has an apparent excess positive charge. This is because the majority charge carriers in P-type material are holes. Holes are the absence of electrons (negative charge), so they act like positive charges in the material. (See the following question.)

E6A04 What is the name given to an impurity atom that adds holes to a semiconductor crystal structure?

A. Insulator impurity
B. N-type impurity
C. Acceptor impurity
D. Donor impurity

C A hole represents a missing electron in the crystal structure. Holes easily accept free electrons to replace the missing electron. For that reason an impurity atom that adds holes to a semiconductor crystal structure is called an acceptor impurity.

E6A05 What is the alpha of a bipolar junction transistor?

A. The change of collector current with respect to base current
B. The change of base current with respect to collector current
C. The change of collector current with respect to emitter current
D. The change of collector current with respect to gate current

C Alpha (α) is the ratio of collector current to emitter current. The smaller the base current, the closer the collector current comes to being equal to that of the emitter, and the closer alpha comes to being 1.

E6A06 What is meant by the beta of a bipolar junction transistor?

A. The frequency at which the current gain is reduced to 1
B. The change in collector current with respect to base current
C. The breakdown voltage of the base to collector junction
D. The switching speed of the transistor

B Beta (β) is the ratio of collector current to base current. It is a measure of the transistor's current gain.

E6A07   In Figure E6-1, what is the schematic symbol for a PNP transistor?

A. 1
B. 2
C. 4
D. 5

A   The PNP transistor symbol is shown at 1 in the drawing. Here's a memory trick. Look at the arrow and remember that to make soup, you put the **P**eas i**N** the **P**ot or, another way to remember; **P**oint i**N** **P**lease.

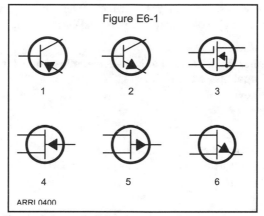

Figure E6-1 — Use this drawing for question E6A07.

E6A08   What term indicates the frequency at which a transistor grounded base current gain has decreased to 0.7 of the gain obtainable at 1 kHz?

A. Corner frequency
B. Alpha rejection frequency
C. Beta cutoff frequency
D. Alpha cutoff frequency

D   The alpha cutoff frequency of a transistor is the frequency at which a transistor's current gain in the grounded-base configuration has decreased to 0.7 of the gain obtainable at 1 kHz.

E6A09   What is a depletion-mode FET?

A. An FET that exhibits a current flow between source and drain when no gate voltage is applied
B. An FET that has no current flow between source and drain when no gate voltage is applied
C. An FET without a channel so no current flows with zero gate voltage
D. An FET without a channel so maximum gate current flows

A   A depletion-mode FET has a channel that passes current from the source to the drain when there is no bias voltage applied between the gate and source. In operation, the gate of a depletion-mode FET is reversed biased (negative gate-to-source voltage). When the reverse bias is applied to the gate, the channel is depleted of charge carriers, and current decreases.

E6A10   In Figure E6-2, what is the schematic symbol for an N-channel dual-gate MOSFET?

A. 2
B. 4
C. 5
D. 6

B   Symbol 4 is a dual-gate N-channel MOSFET. The arrow points iN to the N-channel.

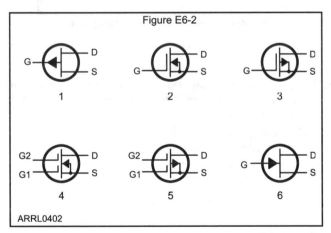

Figure E6-2 — Use this drawing for questions E6A10 and E6A11.

E6A11   In Figure E6-2, what is the schematic symbol for a
P-channel junction FET?

A. 1
B. 2
C. 3
D. 6

A   In the figure, Symbol 1 is a P-channel junction FET. The arrow Points
out of the P-type channel.

E6A12   Why do many MOSFET devices have built-in gate-protective
Zener diodes?

A. To provide a voltage reference for the correct amount of reverse-bias
gate voltage
B. To protect the substrate from excessive voltages
C. To keep the gate voltage within specifications and prevent the device
from overheating
D. To reduce the chance of the gate insulation being punctured by static
discharges or excessive voltages

D   Nearly all the MOSFETs manufactured today have built-in gate-
protective Zener diodes. Without this protection, the gate insulation can be
perforated easily by small static charges on your hand or by the application of
excessive voltages to the device.

E6A13   What do the initials CMOS stand for?

A. Common mode oscillating system
B. Complementary mica-oxide silicon
C. Complementary metal-oxide semiconductor
D. Complementary metal-oxide substrate

C   Sometimes both P- and N-channel MOSFETs are placed on the same
wafer. The resulting transistor arrays can be interconnected on the wafer and
are designed to perform a variety of special functions. This construction is
called complementary metal-oxide semiconductor (CMOS) because the
P- and N-channel transistors operate in complementary ways.

E6A14 How does DC input impedance at the gate of a field-effect transistor compare with the DC input impedance of a bipolar transistor?

A. They cannot be compared without first knowing the supply voltage
B. An FET has low input impedance; a bipolar transistor has high input impedance
C. An FET has high input impedance; a bipolar transistor has low input impedance
D. The input impedance of FETs and bipolar transistors is the same

C Their modes of operation are different and so are their input impedances. The FET has a high input impedance. By contrast, the bipolar transistor has a low input impedance.

E6A15 What two elements widely used in semiconductor devices exhibit both metallic and nonmetallic characteristics?

A. Silicon and gold
B. Silicon and germanium
C. Galena and germanium
D. Galena and bismuth

B Silicon and germanium are the elements most commonly used to make semiconductor materials. These elements exhibit both metallic and nonmetallic characteristics, and their atoms arrange themselves into crystals. Manufacturers add other atoms to these crystals to form the materials that are used to make semiconductor devices.

E6A16 What type of semiconductor material contains fewer free electrons than pure germanium or silicon crystals?

A. N-type
B. P-type
C. Superconductor-type
D. Bipolar-type

B See the discussion for E6A04. A missing electron is the same thing as a hole.

E6A17 **What are the majority charge carriers in N-type semiconductor material?**

A. Holes
B. Free electrons
C. Free protons
D. Free neutrons

B The crystal atoms and the impurity atoms that comprise the material both have an equal number of protons and electrons, but when they are combined into the crystal structure, there are electrons that are not bound to a fixed position in the crystal. These represent free electrons, which are the majority carrier in N-type material.

E6A18 **What are the names of the three terminals of a field-effect transistor?**

A. Gate 1, gate 2, drain
B. Emitter, base, collector
C. Emitter, base 1, base 2
D. Gate, drain, source

D The gate of an FET is the control terminal or element. The source and drain are the two ends of the conducting channel. Current flow through the channel is controlled by voltage applied between the gate and source.

E6B Semiconductor diodes

E6B01 **What is the principal characteristic of a Zener diode?**

A. A constant current under conditions of varying voltage
B. A constant voltage under conditions of varying current
C. A negative resistance region
D. An internal capacitance that varies with the applied voltage

B Zener diodes operate with current flowing in the reverse direction from regular diodes. They are specially manufactured to exhibit a constant voltage drop over a wide range of currents. They have sometimes been called a "poor man's voltage regulator" because of this ability to maintain a constant voltage.

E6B02   What is the principal characteristic of a tunnel diode?

   A. A high forward resistance
   B. A very high PIV
   C. A negative resistance region
   D. A high forward current rating

C   A tunnel diode is a special type of device that has no rectifying properties. When properly biased, it possesses an unusual characteristic: negative resistance. Negative resistance means that when the voltage across the diode increases, the current decreases. This property makes the tunnel diode capable of amplification and oscillation.

E6B03   What is an important characteristic of a Schottky Barrier diode as compared to an ordinary silicon diode when used as a power supply rectifier?

   A. Much higher reverse voltage breakdown
   B. Controlled reverse avalanche voltage
   C. Enhanced carrier retention time
   D. Less forward voltage drop

D   The Schottky Barrier is a special metal-semiconductor junction. A diode made with a Schottky Barrier as opposed to the usual P-N junction has lower forward voltage drop and can rectify high frequency signals more effectively.

E6B04   What special type of diode is capable of both amplification and oscillation?

   A. Point contact
   B. Zener
   C. Tunnel
   D. Junction

C   Because of the negative resistance characteristics of tunnel diodes, they can be used to make both amplifiers and oscillators.

E6B05 What type of semiconductor device varies its internal capacitance as the voltage applied to its terminals varies?

  A. Varactor diode
  B. Tunnel diode
  C. Silicon-controlled rectifier
  D. Zener diode

A Junction diodes exhibit an appreciable internal capacitance due to the regions of charge separated by the narrow depletion layer. It is possible to change the internal capacitance of a diode by varying the amount of reverse bias applied to it. Manufacturers have designed varactor diodes to take advantage of this property.

E6B06 In Figure E6-3, what is the schematic symbol for a varactor diode?

  A. 8
  B. 6
  C. 2
  D. 1

D In the drawing, symbol 1 is the schematic symbol for a varactor diode

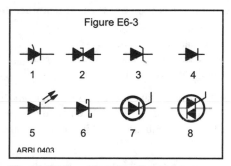

Figure E6-3

ARRL 0403

Figure E6-3 — Use this figure for questions E6B06 and E6B11.

E6B07 What is a common use of a hot-carrier diode?

  A. As balanced mixers in FM generation
  B. As a variable capacitance in an automatic frequency control circuit
  C. As a constant voltage reference in a power supply
  D. As a VHF / UHF mixer or detector

D Hot-carrier diodes have low internal capacitance and good high-frequency characteristics. They are often used in mixers and detectors at VHF and UHF.

E6B08 What limits the maximum forward current rating in a junction diode?

A. Peak inverse voltage
B. Junction temperature
C. Forward voltage
D. Back EMF

B Junction temperature limits the maximum forward current rating in a junction diode. As current flow through a junction diode increases, the junction temperature will rise. If too much current flows, the junction gets too hot and the diode is destroyed.

E6B09 Which of the following describes a type of semiconductor diode?

A. Metal-semiconductor junction
B. Electrolytic rectifier
C. CMOS-field effect
D. Thermionic emission diode

A A metal-semiconductor junction is called a Schottky Barrier. This type of diode is used as a high-speed rectifier.

E6B10 What is a common use for point contact diodes?

A. As a constant current source
B. As a constant voltage source
C. As an RF detector
D. As a high voltage rectifier

C Point-contact diodes have much less internal capacitance than PN-junction diodes. This means that point-contact diodes are better suited for RF applications. They are frequently used in RF detection circuits.

E6B11 In Figure E6-3, what is the schematic symbol for a light-emitting diode?

A. 1
B. 5
C. 6
D. 7

B The schematic symbol for a light-emitting diode is shown in the drawing at number 5.

E6B12 How are junction diodes rated?

A. Maximum forward current and capacitance
B. Maximum reverse current and PIV
C. Maximum reverse current and capacitance
D. Maximum forward current and PIV

D Junction diodes are rated by their ability to handle reverse voltage or PIV (peak inverse voltage). They are also rated by the maximum forward current that they can handle.

E6B13 What is one common use for PIN diodes?

A. As a constant current source
B. As a constant voltage source
C. As an RF switch
D. As a high voltage rectifier

C PIN diodes have regions on each end that are doped to create P-type and N-type material. In between the P and N regions they have an undoped, or intrinsic (I) region. These diodes are particularly useful as RF switches. If you start building RF hardware, at some point you will probably encounter a PIN diode being used as an RF switch.

E6B14 What type of bias is required for an LED to produce luminescence?

A. Reverse bias
B. Forward bias
C. Zero bias
D. Inductive bias

B The LED gives off light when the junction is forward biased and current flows through it.

E6C Integrated circuits: TTL digital integrated circuits; CMOS digital integrated circuits; gates

E6C01 What is the recommended power supply voltage for TTL series integrated circuits?

A. 12 volts
B. 1.5 volts
C. 5 volts
D. 13.6 volts

C The supply voltage can vary between 4.7 and 5.3 V, but 5 V is the preferred voltage for TTL (7400 family) ICs. Other logic families may use other voltages, such as 3.3 V which is common in the logic families used in personal computers.

E6C02 What logic state do the inputs of a TTL device assume if they are left open?

A. A logic-high state
B. A logic-low state
C. The device becomes randomized and will not provide consistent high or low-logic states
D. Open inputs on a TTL device are ignored

A TTL inputs that are left open (disconnected), or allowed to "float," will assume a logic-one or high state. That's a result of their internal construction.

E6C03 What level of input voltage is a logic "high" in a TTL device operating with a positive 5-volt power supply?

A. 2.0 to 5.5 volts
B. 1.5 to 3.0 volts
C. 1.0 to 1.5 volts
D. −5.0 to −2.0 volts

A TTL devices assign a logic 1 (high) to the range of voltages between 2.0 and 5.5 V.

E6C04 What level of input voltage is a logic "low" in a TTL device operating with a positive 5-volt power-supply?

A. −2.0 to −5.5 volts
B. 2.0 to 5.5 volts
C. 0.0 to 0.8 volts
D. −0.8 to 0.4 volts

C A low TTL level is 0 to 0.8 V.

E6C05 Which of the following is an advantage of CMOS logic devices over TTL devices?

A. Differential output capability
B. Lower distortion
C. Immune to damage from static discharge
D. Lower power consumption

D CMOS logic devices are generally smaller, consume less power, and have a lower cost than other logic families. CMOS devices do *not* have a differential output.

E6C06 Why do CMOS digital integrated circuits have high immunity to noise on the input signal or power supply?

A. Larger bypass capacitors arc uscd in CMOS circuit dcsign
B. The input switching threshold is about two times the power supply voltage
C. The input switching threshold is about one-half the power supply voltage
D. Input signals are stronger

C CMOS logic levels typically are within 0.1 V of the supply levels. In other words, if you use a 9 V supply the high level will be in the range of 8.9 to 9 V. In the same circumstances a low level will be from 0 to 0.1 V. The input switching thrcshold is about one-half the power supply voltage, which in this case is 4.5 V. In this example, noise would have to be greater than about 4.4 V to upset circuit operation. This is a substantially higher margin for noise than other logic families, particularly those operating with lower supply voltages.

E6C07 In Figure E6-5, what is the schematic symbol for an AND gate?

A. 1
B. 2
C. 3
D. 4

A The schematic symbol for an AND gate is shown at 1.

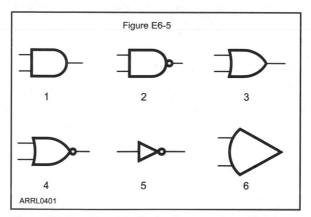

Figure E6-5

1   2   3

4   5   6

ARRL0401

Figure E6-5 — Use this figure for questions E6C07 through E6C11.

E6C08 In Figure E6-5, what is the schematic symbol for a NAND gate?

  A. 1
  B. 2
  C. 3
  D. 4

B The schematic symbol for a NAND gate is symbol number 2 in the drawing.

E6C09 In Figure E6-5, what is the schematic symbol for an OR gate?

  A. 2
  B. 3
  C. 4
  D. 6

B The schematic symbol for an OR gate is shown in the drawing at 3.

E6C10 In Figure E6-5, what is the schematic symbol for a NOR gate?

  A. 1
  B. 2
  C. 3
  D. 4

D The schematic symbol for a NOR gate is shown in the drawing at 4.

E6C11 In Figure E6-5, what is the schematic symbol for the NOT operation (inverter)?

  A. 2
  B. 4
  C. 5
  D. 6

C The schematic symbol for the NOT operation (inverter) is shown at 5.

E6D Optical devices and toroids: vidicon and cathode-ray tube devices; charge-coupled devices (CCDs); liquid crystal displays (LCDs); toroids: permeability, core material, selecting, winding

E6D01 How is the electron beam deflected in a vidicon?

A. By varying the beam voltage
B. By varying the bias voltage on the beam forming grids inside the tube
C. By varying the beam current
D. By varying electromagnetic fields

D A vidicon tube is a relatively simple and inexpensive TV-camera pickup device. An electron beam is used to scan the picture area on the tube face. Horizontal and vertical deflection (scanning) of the electron beam in a vidicon is accomplished with fields generated by coils on the outside of the tube. Varying electromagnetic fields control the beam as it scans the tube face.

E6D02 What is cathode ray tube (CRT) persistence?

A. The time it takes for an image to appear after the electron beam is turned on
B. The relative brightness of the display under varying conditions of ambient light
C. The ability of the display to remain in focus under varying conditions
D. The length of time the image remains on the screen after the beam is turned off

D A given spot of phosphor on the face of a CRT glows for some period of time after the electron beam has moved away. The length of time the phosphor glows after the beam is turned off or moved away is called persistence.

E6D03 If a cathode ray tube (CRT) is designed to operate with an anode voltage of 25,000 volts, what will happen if the anode voltage is increased to 35,000 volts?

A. The image size will decrease
B. The image size will increase
C. The image will become larger and brighter
D. There will be no apparent change

A More is not always better! When the anode voltage is significantly higher than the design voltage, the electrons in the scanning beam will acquire a higher velocity. These higher velocity electrons are deflected less, which means the image size decreases. Another effect of the higher energy electrons is that they may produce X-rays as they collide with the face of the CRT.

E6D04 Exceeding what design rating can cause a cathode ray tube (CRT) to generate X-rays?

A. The heater voltage
B. The anode voltage
C. The operating temperature
D. The operating frequency

B The anode voltage controls the energy of the electrons, which can create X-rays. See the discussion for E6D03.

E6D05 Which of the following is true of a charge-coupled device (CCD)?

A. Its phase shift changes rapidly with frequency
B. It is a CMOS analog-to-digital converter
C. It samples an analog signal and passes it in stages from the input to the output
D. It is used in a battery charger circuit

C A charge-coupled device (CCD) is made from a chain of metal-oxide semiconductor (MOS) capacitors connected by MOSFETs. The first capacitor stores a sample of the analog input signal voltage. When a control pulse biases the MOSFETs to conduct, the first capacitor passes its sampled voltage on to the second capacitor, and it takes a sample by storing a charge from the input. With successive control pulses, each input sample is passed to the next capacitor in the chain. When the MOSFETs are biased off, each capacitor retains its stored charge. This process is sometimes described as a "bucket brigade," because the analog signal is sampled and then passed in stages through the CCD to the output.

E6D06 What function does a charge-coupled device (CCD) serve in a modern video camera?

A. It stores photogenerated charges as signals corresponding to pixels
B. It generates the horizontal pulses needed for electron beam scanning
C. It focuses the light used to produce a pattern of electrical charges corresponding to the image
D. It combines audio and video information to produce a composite RF signal

A A two-dimensional array of CCD elements using light-sensitive materials serves as the image detector in a modern video camera. The charge stored by each capacitor is proportional to the amount of light striking the CCD surface at that point. The charge in each capacitor forms the pixels of the array surface. The signal is shifted out of the CCD array one line of pixels at a time in a pattern that matches a vidicon scan.

E6D07 What is a liquid-crystal display (LCD)?

A. A modern replacement for a quartz crystal oscillator which displays its fundamental frequency
B. A display that uses a crystalline liquid to change the way light is refracted
C. A frequency-determining unit for a transmitter or receiver
D. A display that uses a glowing liquid to remain brightly lit in dim light

B When a voltage is applied across the liquid crystal material in an LCD, it changes the way light is refracted through the material. With no voltage applied, the crystal material is virtually transparent, but with voltage applied, the light is blocked, making the crystal appear black.

E6D08 What material property determines the inductance of a toroidal inductor with a 10-turn winding?

A. Core load current
B. Core resistance
C. Core reactivity
D. Core permeability

D The inductance of a toroidal inductor is determined by the number of turns of wire on the core and by the permeability of the core material. Permeability refers to the strength of a magnetic field in the core as compared to the strength of the field if no core were used. Cores with higher values of permeability will produce larger inductance values for the same number of turns on the coil.

E6D09 What is the usable frequency range of inductors that use toroidal cores, assuming a correct selection of core material for the frequency being used?

A. From a few kHz to no more than 30 MHz
B. From less than 20 Hz to approximately 300 MHz
C. From approximately 1000 Hz to no more than 3000 kHz
D. From about 100 kHz to at least 1000 GHz

B By careful selection of core material, it is possible to produce toroidal inductors that can be used over the range of 20 Hz to around 300 MHz.

E6D10　What is one important reason for using powdered-iron toroids rather than ferrite toroids in an inductor?

A. Powdered-iron toroids generally have greater initial permeabilities
B. Powdered-iron toroids generally have better temperature stability
C. Powdered-iron toroids generally require fewer turns to produce a given inductance value
D. Powdered-iron toroids have the highest power handling capacity

B　The choice of core materials for a particular inductor requires a compromise of characteristics. Ferrite toroids generally have higher permeability values. Powdered-iron cores generally have better temperature stability. Since current flow can cause the toroid core to heat, temperature stability is an important consideration in some applications.

E6D11　What devices are commonly used as VHF and UHF parasitic suppressors at the input and output terminals of transistorized HF amplifiers?

A. Electrolytic capacitors
B. Butterworth filters
C. Ferrite beads
D. Steel-core toroids

C　A ferrite bead is a very small core with a hole designed to slip over a component lead. These are often used as parasitic suppressors at the input and output terminals of transistorized HF amplifiers.

E6D12　What is a primary advantage of using a toroidal core instead of a solenoidal core in an inductor?

A. Toroidal cores contain most of the magnetic field within the core material
B. Toroidal cores make it easier to couple the magnetic energy into other components
C. Toroidal cores exhibit greater hysteresis
D. Toroidal cores have lower Q characteristics

A　A primary advantage of using a toroidal core to wind an inductor rather than a linear core is that nearly all the magnetic field is contained within the core of the toroid. With a linear core, the magnetic field extends into the space surrounding the inductor. A toroidal core inductor can be placed near conducting surfaces and other coils with little effect on its performance.

E6D13 How many turns will be required to produce a 1-mH inductor using a ferrite toroidal core that has an inductance index (A L) value of 523 millihenrys/1000 turns?

A. 2 turns
B. 4 turns
C. 43 turns
D. 229 turns

C   The equation to use for ferrite cores is

$$N = 1000 \sqrt{\frac{L}{A_L}}$$

where L = the inductance in mH; A_L = the inductance index, in mH per 1000 turns; and N = the number of turns.

Plugging the values into the equation gives

$$N = 1000 \sqrt{\frac{L}{A_L}} = 1000 \sqrt{\frac{1}{523}} = 1000 \sqrt{1.91 \times 10^{-3}} = 43.7 \text{ turn}$$

Which is approximately 43 turns.

E6D14 How many turns will be required to produce a 5-microhenry inductor using a powdered-iron toroidal core that has an inductance index (A L) value of 40 microhenrys/100 turns?

A. 35 turns
B. 13 turns
C. 79 turns
D. 141 turns

A   The equation to use for powdered iron cores is

$$N = 100 \sqrt{\frac{L}{A_L}}$$

where L = the inductance in µH; A_L = the inductance index, in µH per 100 turns; and N = the number of turns.

Plug the values into the equation and you get

$$N = 100 \sqrt{\frac{L}{A_L}} = 100 \sqrt{\frac{5}{40}} = 100 \sqrt{0.125} = 35 \text{ turns}$$

E6D15 What type of CRT deflection is better when high-frequency waves are to be displayed on the screen?

A. Electromagnetic
B. Tubular
C. Radar
D. Electrostatic

D While sufficient for television purposes, electromagnetic deflection is not sufficiently precise for measurement purposes. To display high-frequency signals on a lab-type oscilloscope, electrostatic deflection must be used.

E6D16 Which is NOT true of a charge-coupled device (CCD)?

A. It uses a combination of analog and digital circuitry
B. It can be used to make an audio delay line
C. It is commonly used as an analog-to-digital converter
D. It samples and stores analog signals

C A CCD can only work with analog signals. A separate component is required to convert the CCD's output to digital form.

E6D17 What is the principle advantage of liquid-crystal display (LCD) devices over other types of display devices?

A. They consume less power
B. They can display changes instantly
C. They are visible in all light conditions
D. They can be easily interchanged with other display devices

A The principle advantage of LCD devices is that they consume very little power.

E6D18 What is one reason for using ferrite toroids rather than powdered-iron toroids in an inductor?

A. Ferrite toroids generally have lower initial permeabilities
B. Ferrite toroids generally have better temperature stability
C. Ferrite toroids generally require fewer turns to produce a given inductance value
D. Ferrite toroids are easier to use with surface mount technology

C Based on the design equations and the answers to questions E6D13 and E6D14, you can determine the correct answer. Ferrite material generally has a much higher permeability than that of powdered iron. For that reason, ferrite toroids generally require fewer turns to produce a given inductance value.

E6E   Piezoelectric crystals and MMICS: quartz crystals (as used in oscillators and filters); monolithic amplifiers (MMICs)

E6E01   Which of these filter bandwidths would be a good choice for use in a SSB radiotelephone transmitter?

A. 6 kHz at –6 dB
B. 2.4 kHz at –6 dB
C. 500 Hz at –6 dB
D. 15 kHz at –6 dB

B   SSB transmissions require a bit more than 2 kHz of bandwidth. A typical bandwidth is 2.4 kHz at the –6 dB points. This represents the difference between the highest and the lowest voice frequencies that will pass through the filter.

E6E02   Which of these filter bandwidths would be a good choice for use with standard double-sideband AM transmissions?

A. 1 kHz at –6 dB
B. 500 Hz at –6 dB
C. 6 kHz at –6 dB
D. 15 kHz at –6 dB

C   For double-sideband phone emissions, you'll need a bandwidth equal to twice the highest voice frequency that will pass through the filter. That would typically be 6 kHz at –6 dB for a good filter.

E6E03   What is a crystal lattice filter?

A. A power supply filter made with interlaced quartz crystals
B. An audio filter made with four quartz crystals that resonate at 1-kHz intervals
C. A filter with wide bandwidth and shallow skirts made using quartz crystals
D. A filter with narrow bandwidth and steep skirts made using quartz crystals

D   Crystal lattice filters are made using quartz crystals. They are characterized as having a relatively narrow bandwidth and steep skirts.

E6E04   What technique is used to construct low-cost, high-performance crystal ladder filters?

A. Obtain a small quantity of custom-made crystals
B. Choose a crystal with the desired bandwidth and operating frequency to match a desired center frequency
C. Measure crystal bandwidth to ensure at least 20% coupling
D. Measure crystal frequencies and carefully select units with a frequency variation of less than 10% of the desired filter bandwidth

D   If you wish to construct a low-cost, high-performance crystal ladder filter, start with a collection of crystals that have approximately the same frequency. Measure their frequencies using an oscillator circuit. Select crystals with a frequency variation of less than 10% from the desired filter bandwidth. In other words, if you want to build a 500-Hz bandwidth filter, select crystals that have resonant frequencies all within 50 Hz of each other.

E6E05   Which of the following factors has the greatest effect in helping determine the bandwidth and response shape of a crystal ladder filter?

A. The relative frequencies of the individual crystals
B. The DC voltage applied to the quartz crystal
C. The gain of the RF stage preceding the filter
D. The amplitude of the signals passing through the filter

A   The bandwidth and response shape of a crystal ladder filter mostly depends on the relative frequencies of the individual crystals.

E6E06   What is one aspect of the piezoelectric effect?

A. Physical deformation of a crystal by the application of a voltage
B. Mechanical deformation of a crystal by the application of a magnetic field
C. The generation of electrical energy by the application of light
D. Reversed conduction states when a P-N junction is exposed to light

A   The piezoelectric effect is the physical deformation of a material, such as a quartz crystal, when a voltage is applied to it. This physical deformation results in vibrations at a particular frequency and these vibrations can be used to control the operating frequency of a circuit.

E6E07   What is the characteristic impedance of circuits in which almost all MMICs are designed to work?

A. 50 ohms
B. 300 ohms
C. 450 ohms
D. 10 ohms

A   This is the first in a series of several questions about MMIC (monolithic microwave integrated circuit) design. Since MMICs frequently interface with other RF devices and systems, they are designed with a characteristic impedance of 50 Ω for the input and the output.

E6E08   What is the typical noise figure of a monolithic microwave integrated circuit (MMIC) amplifier?

A. Less than 1 dB
B. Approximately 3.5 to 6 dB
C. Approximately 8 to 10 dB
D. More than 20 dB

B   MMICs have good noise figure performance. Typical performance is in the neighborhood of 3 to 6 dB.

E6E09   What type of amplifier device consists of a small pill-type package with an input lead, an output lead and 2 ground leads?

A. A junction field-effect transistor (JFET)
B. An operational amplifier integrated circuit (OAIC)
C. An indium arsenide integrated circuit (IAIC)
D. A monolithic microwave integrated circuit (MMIC)

D   The MMIC is unlike most other ICs that you may be familiar with. These ICs are quite small, often classified as "pill sized" devices, perhaps because they look like a small disc-shaped pill with four leads coming out of the device at 90° to each other. There is an input lead, an output lead and two ground leads on a typical MMIC.

E6E10 What typical construction technique is used when building an amplifier for the microwave bands containing a monolithic microwave integrated circuit (MMIC)?

A. Ground-plane "ugly" construction
B. Microstrip construction
C. Point-to-point construction
D. Wave-soldering construction

B MMICs are surface mount devices and MMIC construction typically uses microstrip techniques. Double-sided circuit board material is used, and one side serves to form a ground plane for the circuit. Circuit traces form sections of feed line. The line widths, along with the circuit-board thickness and dielectric constant of the insulating material determine the characteristic impedance, which is normally 50 Ω.

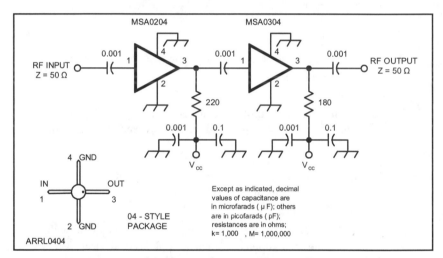

Figure E6E11 — This circuit illustrates how MMICs are connected in a practical circuit. Power is supplied to the output pin through decoupling circuits and dc blocking capacitors allow the output signal from one MMIC to be connected to the input of the next.

E6E11 How is the operating bias voltage normally supplied to the most common type of monolithic microwave integrated circuit (MMIC)?

A. Through a resistor and/or RF choke connected to the amplifier output lead
B. MMICs require no operating bias
C. Through a capacitor and RF choke connected to the amplifier input lead
D. Directly to the bias-voltage (VCC IN) lead

A The operating bias voltage is normally supplied to a four-lead MMIC through a decoupling resistor or RF choke connected to the amplifier output lead. This technique is illustrated in Figure E6E11.

E6E12 What supply voltage do monolithic microwave integrated circuits (MMIC) amplifiers typically require?

A. 1 volt DC
B. 12 volts DC
C. 20 volts DC
D. 120 volts DC

B MMIC amplifiers typically require 12 volts dc.

E6E13 What is the most common package for inexpensive monolithic microwave integrated circuit (MMIC) amplifiers?

A. Beryllium oxide packages
B. Glass packages
C. Plastic packages
D. Ceramic packages

C Integrated circuits typically come in plastic and ceramic packages. Of those two choices, plastic is the less expensive, and that's what they use for inexpensive MMIC amplifiers.

E6F Optical components and power systems: photo conductive principles and effects, photo voltaic systems, optical couplers, optical sensors, and optoisolators

E6F01 What is photoconductivity?

A. The conversion of photon energy to electromotive energy
B. The increased conductivity of an illuminated semiconductor
C. The conversion of electromotive energy to photon energy
D. The decreased conductivity of an illuminated semiconductor

B The total conductance of a material may increase and its resistance decrease when light shines on the surface. This is called the photoconductive effect, or photoconductivity.

E6F02 What happens to the conductivity of a photoconductive material when light shines on it?

A. It increases
B. It decreases
C. It stays the same
D. It becomes unstable

A The conductance of a photoconductive material increases when light shines on it.

E6F03 What is the most common configuration for an optocoupler?

A. A lens and a photomultiplier
B. A frequency modulated helium-neon laser
C. An amplitude modulated helium-neon laser
D. An LED and a phototransistor

D An optoisolator or optocoupler is an LED and a phototransistor in a single IC package. Light from the LED shines on the phototransistor, allowing current to flow in the output circuit.

E6F04 Which of the following is an optoisolator?

- A. An LED and a phototransistor
- B. A P-N junction that develops an excess positive charge when exposed to light
- C. An LED and a capacitor
- D. A P-N junction that develops an excess negative charge when exposed to light

A Optoisolators and optocouplers are basically the same type of device. In addition to the coupling of the LED and phototransistor, there is a very high impedance between the input and the output of these devices. This means there is no current between the input and output circuits.

E6F05 What is an optical shaft encoder?

- A. An array of neon or LED indicators whose light transmission path is controlled by a rotating wheel
- B. An array of optocouplers whose light transmission path is controlled by a rotating wheel
- C. An array of neon or LED indicators mounted on a rotating wheel in a coded pattern
- D. An array of optocouplers mounted on a rotating wheel in a coded pattern

B An optical shaft encoder usually consists of two pairs of emitters and detectors. A plastic disc with alternating clear and black radial bands rotates through a gap between the emitters and detectors. By using two emitters and two detectors, a microprocessor can detect the rotation direction and speed of the wheel. Modern transceivers use a system like this to control the frequency of a synthesized VFO.

E6F06 What characteristic of a crystalline solid will photoconductivity change?

- A. The capacitance
- B. The inductance
- C. The specific gravity
- D. The resistance

D You might have expected to find conductance among the choices, but it's not there. The correct answer is resistance, which is the reciprocal of conductance. In other words, if you change one, you change the other — but in the opposite way.

E6F07 Which material will exhibit the greatest photoconductive effect when illuminated by visible light?

A. Potassium nitrate
B. Lead sulfide
C. Cadmium sulfide
D. Sodium chloride

C Common, inexpensive photodetectors for visible light are typically made from cadmium sulfide.

E6F08 Which material will exhibit the greatest photoconductive effect when illuminated by infrared light?

A. Potassium nitrate
B. Lead sulfide
C. Cadmium sulfide
D. Sodium chloride

B Lead sulfide is the best choice for sensitivity to infrared light.

E6F09 Which of the following materials is affected the most by photoconductivity?

A. A crystalline semiconductor
B. An ordinary metal
C. A heavy metal
D. A liquid semiconductor

A Crystalline semiconductors work best. Metals can be affected by photoconductivity but not as much as semiconductors. Liquid semiconductors are affected less than crystalline semiconductors.

E6F10 What characteristic of optoisolators is often used in power supplies?

A. They have low impedance between the light source and the phototransistor
B. They have very high impedance between the light source and the phototransistor
C. They have low impedance between the light source and the LED
D. They have very high impedance between the light source and the LED

B Optoisolators have a very high impedance between the light source (input) and the phototransistor (output). That characteristic makes these devices particularly useful for interfacing circuits with different voltages or ground references.

E6F11   What characteristic of optoisolators makes them suitable for use with a triac to form the solid-state equivalent of a mechanical relay for a 120 V AC household circuit?

A. Optoisolators provide a low impedance link between a control circuit and a power circuit
B. Optoisolators provide impedance matching between the control circuit and power circuit
C. Optoisolators provide a very high degree of electrical isolation between a control circuit and a power circuit
D. Optoisolators eliminate (isolate) the effects of reflected light in the control circuit

C   They are called isolators because they have a very high impedance between the input and the output. This provides a high degree of electrical isolation between the controlling circuit and the controlled circuit.

E6F12   Which of the following types of photovoltaic cell has the highest efficiency?

A. Silicon
B. Silver iodide
C. Selenium
D. Gallium arsenide

D   Crystalline gallium arsenide converts more of the illumination to electricity than other semiconductor material. Photovoltaic technology is making rapid strides, however, and other materials may be more found with higher conversion efficiencies.

E6F13   What is the most common type of photovoltaic cell used for electrical power generation?

A. Selenium
B. Silicon
C. Cadmium Sulfide
D. Copper oxide

B   Most photovoltaic or solar cells are made from crystalline silicon, the same material used in most transistors. Even though it does not have the highest efficiency, it is inexpensive because of its wide use. As mentioned in the previous question, technology is making rapid strides and other materials with higher conversion efficiencies may eventually be used.

E6F14 Which of the following is the approximate open-circuit voltage produced by a fully-illuminated silicon photovoltaic cell?

A. 0.1 V
B. 0.5 V
C. 1.5 V
D. 12 V

B Open-circuit means the output voltage with no load connected. In full sunlight, a typical silicon solar cell will develop 0.5 V at its output terminals.

E6F15 What absorbs the energy from light falling on a photovoltaic cell?

A. Protons
B. Photons
C. Electrons
D. Holes

C When an electron near a semiconductor solar cell's depletion region absorbs a photon, the additional energy enables it to travel across the junction and combine with a hole to generate electricity.

Practical Circuits

There will be eight questions on your Extra class examination from the Practical Circuits subelement. These eight questions will be taken from the eight groups of questions labeled E7A through E7H.

E7A Digital circuits: digital circuit principles and logic circuits: classes of logic elements; positive and negative logic; frequency dividers; truth tables

E7A01 What is a bistable circuit?
A. An "AND" gate
B. An "OR" gate
C. A flip-flop
D. A clock

C   "Bistable" means that the circuit has two (bi) stable states, such as on or off. "Bistable multivibrator" is just another name for a digital logic flip-flop. For that reason, a flip-flop can be used to store one bit of information.

E7A02 How many output level changes are obtained for every two trigger pulses applied to the input of a "T" flip-flop circuit?
A. None
B. One
C. Two
D. Four

C   In the case of a toggle, or T-type, flip-flop, each input pulse changes the output state. Thus, two pulses produce two output changes.

E7A03 Which of the following can divide the frequency of a pulse train by 2?

A. An XOR gate
B. A flip-flop
C. An OR gate
D. A multiplexer

B A flip-flop, also known as a bistable multivibrator (see the discussion for E7A01), can be used as a frequency divider. Since the output state changes once for each input pulse, it takes two pulses to complete the output cycle from one binary state, to the opposite state, then back to the original state.

E7A04 How many flip-flops are required to divide a signal frequency by 4?

A. 1
B. 2
C. 4
D. 8

B Each flip-flop has the ability to divide a signal's frequency by 2, so the answer to this question is 4/2 = 2.

E7A05 Which of the following is a circuit that continuously alternates between two unstable states without an external clock?

A. Monostable multivibrator
B. J-K Flip-Flop
C. T Flip-Flop
D. Astable Multivibrator

D If you know your language, you will recognize the prefix "*a*" as meaning "*not*" so an astable (not stable) multivibrator alternates between two unstable states. Ordinarily, an unstable circuit is undesirable, but the astable multivibrator is useful as a digital oscillator to generate clock or other timing signals.

E7A06 What is a characteristic of a monostable multivibrator?

    A. It switches momentarily to the opposite binary state and then returns, after a set time, to its original state

    B. It is a clock that produces a continuous square wave oscillating between 1 and 0

    C. It stores one bit of data in either a 0 or 1 state

    D. It maintains a constant output voltage, regardless of variations in the input voltage

A   A monostable multivibrator, meaning "stable in only one state" and also called a "single shot" or "one-shot," switches momentarily to the opposite binary state and then returns after a set time to its original state. The length of the momentary period is determined by circuit component values.

E7A07 What logical operation does an AND gate perform?

    A. It produces a logic "0" at its output only if all inputs are logic "1"

    B. It produces a logic "1" at its output only if all inputs are logic "1"

    C. It produces a logic "1" at its output if only one input is a logic "1"

    D. It produces a logic "1" at its output if all inputs are logic "0"

B   Table E7-1 lists the functions of common two-input logic elements. By looking at the table, which is called a "truth table," you see that an AND gate requires a logic "1" on both inputs to produce a logic "1" at the output. An "N" before the AND or OR means NOT, or a logical inversion, so a NAND is a NOT-AND. The X in XOR stands for "exclusive," meaning "one of but not both."

Table E7-1

Output of Various Logic Gates

Input 1	Input 2	AND	NAND	OR	NOR	XOR
0	0	0	1	0	1	0
0	1	0	1	1	0	1
1	0	0	1	1	0	1
1	1	1	0	1	0	0

E7A08 What logical operation does a NAND gate perform?

    A. It produces a logic "0" at its output only when all inputs are logic "0"

    B. It produces a logic "1" at its output only when all inputs are logic "1"

    C. It produces a logic "0" at its output if some but not all of its inputs are logic "1"

    D. It produces a logic "0" at its output only when all inputs are logic "1"

D   Table E7-1 shows that a NAND (NOT-AND) gate produces a logic "0" at its output only when all inputs are logic "1."

E7A09 What logical operation does an OR gate perform?

    A. It produces a logic "1" at its output if any or all inputs are logic "1"
    B. It produces a logic "0" at its output if all inputs are logic "1"
    C. It only produces a logic "0" at its output when all inputs are logic "1"
    D. It produces a logic "1" at its output if all inputs are logic "0"

A   Table E7-1 shows that an OR gate produces a logic "1" at its output if either or both of its inputs are logic "1."

E7A10 What logical operation does a NOR gate perform?

    A. It produces a logic "0" at its output only if all inputs are logic "0"
    B. It produces a logic "1" at its output only if all inputs are logic "1"
    C. It produces a logic "0" at its output if any or all inputs are logic "1"
    D. It produces a logic "1" at its output only when none of its inputs are logic "0"

C   Table E7-1 shows that the NOR (NOT-OR) gate produces a logic "0" at its output if any input is or all inputs are logic "1."

E7A11 What is a truth table?

    A. A table of logic symbols that indicate the high logic states of an op-amp
    B. A diagram showing logic states when the digital device's output is true
    C. A list of inputs and corresponding outputs for a digital device
    D. A table of logic symbols that indicates the low logic states of an op-amp

C   Table E7-1 is an example of a truth table. A truth table lists input combinations and their corresponding outputs for a digital device or devices. In other words, the truth table shows all of the combinations that are "true" for that device.

E7A12 What is the name for logic which represents a logic "1" as a high voltage?

    A. Reverse Logic
    B. Assertive Logic
    C. Negative logic
    D. Positive Logic

D   In positive logic, a 1 is represented by a high voltage level.

E7A13 What is the name for logic which represents a logic "0" as a high voltage?

A. Reverse Logic
B. Assertive Logic
C. Negative logic
D. Positive Logic

C As you might have guessed, this is the opposite of the previous question. In a negative-logic circuit, a low level is used to represent a logic 1.

E7B Amplifiers: Class of operation; vacuum tube and solid-state circuits; distortion and intermodulation; spurious and parasitic suppression; microwave amplifiers

E7B01 For what portion of a signal cycle does a Class AB amplifier operate?

A. More than 180 degrees but less than 360 degrees
B. Exactly 180 degrees
C. The entire cycle
D. Less than 180 degrees

A The Class AB amplifier conducts for more than 180 degrees (Class B) but less than 360 degrees (Class A).

E7B02 Which class of amplifier, of the types shown, provides the highest efficiency?

A. Class A
B. Class B
C. Class C
D. Class AB

C A Class C amplifier provides the highest efficiency of the classes listed in the question. There are other classes of amplifiers (Class D through Class G) that are based on the same types of circuits as switching power supplies and have very high efficiencies.

E7B03 Where on the load line of a Class A common emitter amplifier would bias normally be set?

A. Approximately half-way between saturation and cutoff
B. Where the load line intersects the voltage axis
C. At a point where the bias resistor equals the load resistor
D. At a point where the load line intersects the zero bias current curve

A A load line describes the possible combinations of output voltage and current. One end of the line is at zero current and full voltage (cutoff). The other lies at full current and zero voltage (saturation). The amplifier circuit's design determines where the operating point falls along that line. Class A power amplifiers must be operated away from the saturation or cutoff points for linear amplification. Once an amplifier reaches saturation or cutoff, it is no longer operating in Class A.

E7B04 What can be done to prevent unwanted oscillations in a power amplifier?

A. Tune the stage for maximum SWR
B. Tune both the input and output for maximum power
C. Install parasitic suppressors and/or neutralize the stage
D. Use a phase inverter in the output filter

C Neutralization is a technique that prevents parasitic oscillations in a power amplifier due to positive feedback within the amplifying devices.

E7B05 Which of the following amplifier types reduces or eliminates even-order harmonics?

A. Push-push
B. Push-pull
C. Class C
D. Class AB

B A push-pull amplifier with two amplifying devices operated in Class B reduces even-order harmonics.

E7B06 Which of the following is a likely result when a Class C rather than a class AB amplifier is used to amplify a single-sideband phone signal?

A. Intermodulation products will be greatly reduced
B. Overall intelligibility will increase
C. Part of the transmitted signal will be inverted
D. The signal may become distorted and occupy excessive bandwidth

D Class C amplifiers are highly nonlinear, and nonlinear amplifiers cause distortion. The distorted waveform contains harmonics and spurious signals (splatter) that occupy a wide bandwidth. SSB requires linear amplification to be received properly and to occupy the minimum bandwidth necessary.

E7B07 How can a vacuum-tube power amplifier be neutralized?

A. By increasing the grid drive
B. By reducing the grid drive
C. By feeding back an out-of-phase component of the output to the input
D. By feeding back an in-phase component of the output to the input

C A certain amount of capacitance exists between the input and output circuits in any active device, tube or transistor. This capacitance creates in-phase (positive) feedback of some of the output signal to the input. This can cause instability and even oscillations. Neutralization uses a negative, or out-of-phase, feedback signal from the output to the input to offset or neutralize this in-phase feedback signal.

E7B08 Which of the following describes how the loading and tuning capacitors are to be adjusted when tuning a vacuum tube RF power amplifier that employs a pi-network output circuit?

A. The loading capacitor is set to maximum capacitance and the tuning capacitor is adjusted for minimum allowable plate current
B. The tuning capacitor is set to maximum capacitance and the loading capacitor is adjusted for minimum plate permissible current
C. The loading capacitor is adjusted to minimum plate current while alternately adjusting the tuning capacitor for maximum allowable plate current
D. The tuning capacitor is adjusted for minimum plate current, while the loading capacitor is adjusted for maximum permissible plate current

D The procedure for tuning a vacuum-tube power amplifier having an output pi-network is to alternately increase the plate current with the loading capacitor and dip the plate current with the tuning capacitor. The goal is to produce the required amount of output power with the minimum amount of plate current.

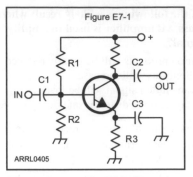

Figure E7-1 — Use this figure to answer questions E7B09 through E7B11.

E7B09   In Figure E7-1, what is the purpose of R1 and R2?

A. Load resistors
B. Fixed bias
C. Self bias
D. Feedback

B   The two resistors form a voltage divider that provides a fixed bias voltage and current for the base.

E7B10   In Figure E7-1, what is the purpose of R3?

A. Fixed bias
B. Emitter bypass
C. Output load resistor
D. Self bias

D   Current through R3 creates a voltage drop across the resistor, which acts to oppose the bias provided by R1 and R2. This stabilizes the transistor's collector current and is called self-bias.

E7B11   What type of circuit is shown in Figure E7-1?

A. Switching voltage regulator
B. Linear voltage regulator
C. Common emitter amplifier
D. Emitter follower amplifier

C   This is a transistor amplifier circuit. Since the emitter is grounded and the output is taken from the collector circuit, it is called a common (or grounded) emitter amplifier.

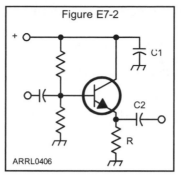

Figure E7-2 — Use this figure to answer questions E7B12 and E7B13.

E7B12 In Figure E7-2, what is the purpose of R?

A. Emitter load
B. Fixed bias
C. Collector load
D. Voltage regulation

A Since the resistor is connected to the emitter, you might guess that it represents the emitter load. If so, you'd be right.

E7B13 In Figure E7-2, what is the purpose of C2?

A. Output coupling
B. Emitter bypass
C. Input coupling
D. Hum filtering

A The capacitor C2 is connected to the output terminal and its job is output coupling. That is, it blocks dc voltage from the circuit while passing ac current to the load.

E7B14 What is one way to prevent thermal runaway in a transistor amplifier?

A. Neutralization
B. Select transistors with high beta
C. Use degenerative emitter feedback
D. All of the above

C Degenerative emitter feedback is another term for self-bias as created by the emitter resistor in question E7B10. The feedback consists of the voltage drop across the emitter. It is degenerative because as the voltage increases from higher emitter current, it acts to reduce the amount of bias and collector current. Degenerative feedback is a type of negative feedback.

Practical CIrcuits   9

E7B15 What is the effect of intermodulation products in a linear power amplifier?

A. Transmission of spurious signals
B. Creation of parasitic oscillations
C. Low efficiency
D. All of the above

A The mixing products that result from intermodulation create additional sidebands on frequencies near the desired signal. These spurious signals interfere with communications on these frequencies.

E7B16 Why are third-order intermodulation distortion products of particular concern in linear power amplifiers?

A. Because they are relatively close in frequency to the desired signal
B. Because they are relatively far in frequency from the desired signal
C. Because they invert the sidebands causing distortion
D. Because they maintain the sidebands, thus causing multiple duplicate signals

A Third-order intermodulation products from nonlinear amplification of complex signals such as SSB phone signals are near the original frequency. For example, a two-tone SSB signal with components f_1 and f_2 generates third-order intermodulation products at $2f_1 \pm f_2$ and $2f_2 \pm f_1$. These are close enough to the original signals to interfere with communications in the same band.

E7B17 Which of the following is a characteristic of a grounded-grid amplifier?

A. High power gain
B. High filament voltage
C. Low input impedance
D. Low bandwidth

C Although other types of amplifier circuits have higher power gain, grounded-grid amplifiers are popular because their low input impedance is easy to match to the 50 Ω of transmitter output circuits. They also have a wide operating bandwidth of useful power gain.

E7B18 What is a klystron?

A. A high speed multivibrator
B. An electron-coupled oscillator utilizing a pentode vacuum tube
C. An oscillator utilizing ceramic elements to achieve stability
D. A VHF, UHF, or microwave vacuum tube that uses velocity modulation

D The klystron is a type of traveling-wave amplifier in which the electrons are formed into a beam that travels inside a cylindrical cavity called a drift tube. The tube has small gaps spaced at precise intervals. RF voltages are applied to the gaps and the resulting electric fields cause the electrons to accelerate or decelerate. As they travel through the drift tube, the repeated acceleration and deceleration causes the density of the electrons in the beam to vary in a regular pattern, like waves on the surface of a lake. When the beam of electrons reaches the anode, the variations in beam density become variations in current in the anode circuit. This technique called velocity modulation.

E7B19 What is a parametric amplifier?

A. A type of bipolar operational amplifier with excellent linearity derived from use of very high voltage on the collector
B. A low-noise VHF or UHF amplifier relying on varying reactance for amplification
C. A high power amplifier for HF application utilizing the Miller effect to increase gain
D. An audio push-pull amplifier using silicon carbide transistors for extremely low noise

B The parametric amplifier amplifies signals by using a diode as a small variable capacitor and a separate oscillator signal that "pumps" the diode. The change in capacitance creates a change in voltage greater than that of the input signal. The resulting amplification is low-noise because the only contributor to noise in the output is that of the oscillator.

E7B20 Which of the following devices is generally best suited for UHF or microwave power amplifier applications?

A. FET
B. Nuvistor
C. Silicon Controlled Rectifier
D. Triac

A Semiconductor power amplifiers at UHF and higher frequencies usually use FETs (Field Effect Transistors) as the amplifying devices because they are more efficient at these frequencies than bipolar transistors. FETs can also provide power gain at higher frequencies than bipolar transistors.

E7C Filters and matching networks: filters and impedance matching networks: types of networks; types of filters; filter applications; filter characteristics; impedance matching; DSP filtering

E7C01 How are the capacitors and inductors of a low-pass filter Pi-network arranged between the network's input and output?

A. Two inductors are in series between the input and output and a capacitor is connected between the two inductors and ground
B. Two capacitors are in series between the input and output and an inductor is connected between the two capacitors and ground
C. An inductor is in parallel with the input, another inductor is in parallel with the output, and a capacitor is in series between the two
D. A capacitor is in parallel with the input, another capacitor is in parallel with the output, and an inductor is in series between the two

D The correct configuration should look like the Greek letter π (pi). Since the question is looking for the low pass filter, we want the one with the inductor in the middle. This configuration is low-pass because capacitors pass higher frequencies and block low frequencies while inductors pass low frequencies and block high frequencies. This shunts the high frequencies to ground and blocks them from passing to the output. In the high-pass filter pi-network, the low frequencies are shunted to ground while the high frequencies pass through.

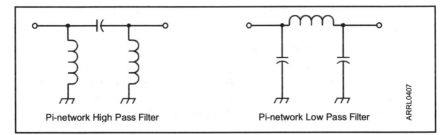

Pi-network High Pass Filter   Pi-network Low Pass Filter

ARRL0407

Figure E7C01 — Pi-network filters have one series and two shunt elements.

E7C02 A T-network with series capacitors and a parallel (shunt) inductor has which of the following properties?

A. It transforms impedance and is a low-pass filter
B. It transforms reactance and is a low-pass filter
C. It transforms impedance and is a high-pass filter
D. It transforms reactance and is a narrow bandwidth notch filter

C   The T-network transforms impedances as a pair of back-to-back L networks. It is a high pass filter because the inductor shunts the low frequencies to ground while the capacitors allow the high frequencies to pass through.

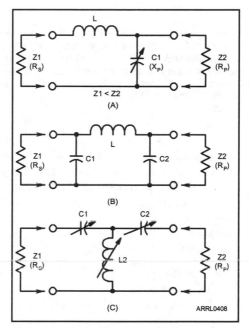

Figure E7C02 — Three types of impedance matching networks. (A) is an L network, (B) is a Pi network, and (C) is a T network. A and B are shown in a low-pass configuration. The T network at C acts as a high-pass filter.

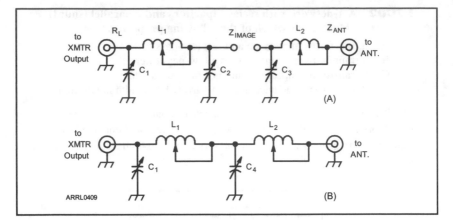

Figure E7C03 — In circuit A, a pi-L network uses a pi network to transform the transmitter output impedance (R_L) to the image impedance (Z_{IMAGE}). An L network then transforms the image impedance to the antenna impedance, Z_{ANT}. Because C_2 and C_3 are in parallel, they are combined into a single capacitor (C_4) as shown in Circuit B.

E7C03 What advantage does a Pi-L-network have over a Pi-network for impedance matching between the final amplifier of a vacuum-tube type transmitter and an antenna?

A. Greater harmonic suppression
B. Higher efficiency
C. Lower losses
D. Greater transformation range

A The additional inductor of the pi-L network increases high frequency rejection at the cost of additional losses, reduced efficiency, and a somewhat reduced range of impedances that can be transformed.

E7C04 How does a network transform a complex impedance to a resistive impedance?

A. It introduces negative resistance to cancel the resistive part of an impedance
B. It introduces transconductance to cancel the reactive part of an impedance
C. It cancels the reactive part of an impedance and transforms the resistive part to the desired value
D. Network resistances are substituted for load resistances

C Impedance matching networks transform one impedance to another by changing the ratios of voltage and current (impedance) and by changing the phase relationship between the voltage and current (reactance).

E7C05 Which filter type is described as having ripple in the passband and a sharp cutoff?

A. A Butterworth filter
B. An active LC filter
C. A passive op-amp filter
D. A Chebyshev filter

D The Chebyshev filter design equations result in a filter with amplitude response variations (ripple) in the passband. This tradeoff is made to obtain a sharp cutoff.

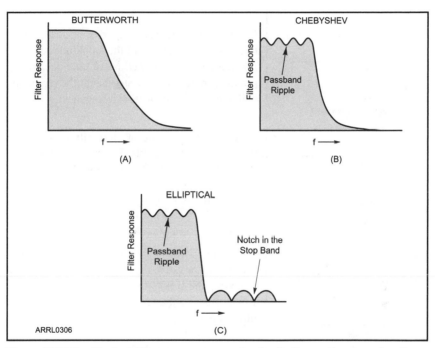

Figure E7C05 — Typical filter response curves for Butterworth or maximally-flat (A), Chebyshev (B), and Elliptical (C) designs.

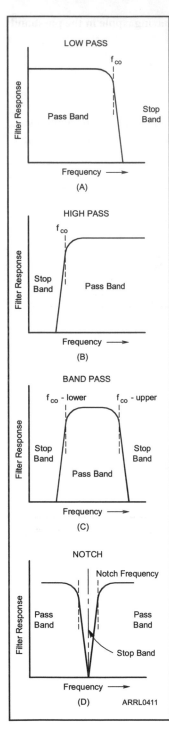

LOW PASS

f_{co}

Filter Response

Pass Band

Stop Band

Frequency ⟶

(A)

HIGH PASS

f_{co}

Filter Response

Stop Band

Pass Band

Frequency ⟶

(B)

BAND PASS

f_{co} - lower f_{co} - upper

Filter Response

Stop Band

Stop Band

Pass Band

Frequency ⟶

(C)

NOTCH

Notch Frequency

Filter Response

Pass Band

Pass Band

Stop Band

Frequency ⟶

(D)   ARRL0411

E7C06 **What are the distinguishing features of an elliptical filter?**

A. Gradual passband rolloff with minimal stop-band ripple
B. Extremely flat response over its passband, with gradually rounded stop-band corners
C. Extremely sharp cutoff, with one or more infinitely deep notches in the stop band
D. Gradual passband rolloff with extreme stop-band ripple

C Like the Chebyshev filters, elliptical filters trade away the flatness of the amplitude response for sharper cutoff. The elliptical filter design also produces notches in the stop band (the region in which the filter attenuates frequencies).

E7C07 **What kind of audio filter would you use to attenuate an interfering carrier signal while receiving an SSB transmission?**

A. A band-pass filter
B. A notch filter
C. A Pi-network filter
D. An all-pass filter

B A notch filter (or band-stop filter) attenuates a very narrow range of frequencies, just right for removing the single tone of interfering carriers while affecting the remainder of the signal as little as possible.

Figure E7C07 — Ideal filter response curves for low-pass, high-pass, band-pass, and notch filters.

E7C08 What kind of digital signal processing audio filter might be used to remove unwanted noise from a received SSB signal?

A. An adaptive filter
B. A crystal-lattice filter
C. A Hilbert-transform filter
D. A phase-inverting filter

A   Because the DSP filter is created by software, it can sense the conditions at its input and adapt its operation accordingly. This allows a DSP adaptive filter to remove all manner of interfering noise from voice and CW signals.

E7C09 What type of digital signal processing filter might be used in generating an SSB signal?

A. An adaptive filter
B. A notch filter
C. A Hilbert-transform filter
D. An elliptical filter

C   The Hilbert-transform filter is a special type of filter that is straightforward to implement with DSP. The filter implements a set of mathematical equations that turn a carrier and an information signal into an SSB signal without a carrier or unwanted sideband.

E7C10 Which of the following filters would be the best choice for use in a 2-meter repeater duplexer?

A. A crystal filter
B. A cavity filter
C. A DSP filter
D. An L-C filter

B   Cavity filters are resonant enclosures with extremely low losses and thus high Q and narrow bandwidths. Cavity filters can be configured to pass or remove signals, making them especially useful for repeaters where the transmit and receive frequencies are quite close together.

E7C11 Which of the following is the common name for a filter network which is equivalent to two L networks back-to-back?

A. Pi-L
B. Cascode
C. Omega
D. Pi

D The pi network takes its name from its resemblance to the Greek letter π. A pi network consists of one inductor and two capacitors or two inductors and one capacitor. The T network is another circuit that can be thought of as a pair of back-to-back L networks.

E7C12 What is a Pi-L network, as used when matching a vacuum-tube final amplifier to a 50-ohm unbalanced output?

A. A Phase Inverter Load network
B. A network consisting of two series inductors and two shunt capacitors
C. A network with only three discrete parts
D. A matching network in which all components are isolated from ground

B A pi-L-network is essentially a pi-network with an additional series element. The pi-L network most common in amateur consists of two inductors as the series components and two capacitors as the shunt elements.

E7C13 What is one advantage of a Pi matching network over an L matching network?

A. Q of Pi networks can be varied depending on the component values chosen
B. L networks can not perform impedance transformation
C. Pi networks have fewer components
D. Pi networks are designed for balanced input and output

A Pi networks transform impedances in two steps, one by the input pair of components and one by the output pair of components. By choosing the transformation ratios of each stage, the Q of the overall circuit can also be adjusted.

E7C14 Which of these modes is most affected by non-linear phase response in a receiver IF filter?

A. Meteor Scatter
B. Single-Sideband Voice
C. Digital
D. Video

C   Digital communications often use relative phase to encode the data symbols and so those modes are highly dependent on the phase response of the receiver filters.

E7D   Power supplies and voltage regulators

E7D01 What is one characteristic of a linear electronic voltage regulator?

A. It has a ramp voltage as its output
B. It eliminates the need for a pass transistor
C. The control element duty cycle is proportional to the line or load conditions
D. The conduction of a control element is varied to maintain a constant output voltage

D   A linear voltage regulator varies the conductance of a control element (such as a transistor) to maintain a constant output voltage to the load. Because this is a linear circuit, the control is continuous. There are no switching elements used in this type of regulator.

E7D02 What is one characteristic of a switching electronic voltage regulator?

A. The resistance of a control element is varied in direct proportion to the line voltage or load current
B. It is generally less efficient than a linear regulator
C. The control device's duty cycle is controlled to produce a constant average output voltage
D. It gives a ramp voltage at its output

C   A switching regulator, in contrast to a linear regulator, has switching elements (such as transistors) that control the regulator output voltage. In switching regulators the control device is switched on and off electronically with the duty cycle automatically adjusted to maintain a constant average output voltage. Switching frequencies of several kilohertz and higher are used to avoid the need for extensive filtering to remove ripple at the switching frequency from the dc output.

E7D03 What device is typically used as a stable reference voltage in a linear voltage regulator?

A. A Zener diode
B. A tunnel diode
C. An SCR
D. A varactor diode

A   A Zener diode can provide a stable voltage reference and is frequently used for that purpose in linear voltage regulators.

E7D04 Which of the following types of linear regulator makes the most efficient use of the primary power source?

A. A constant current source
B. A series regulator
C. A shunt regulator
D. A shunt current source

B   The series regulator draws current from the primary power source in proportion to the load. In other words, it only draws power from the primary source when that power is needed.

E7D05 Which of the following types of linear voltage regulator places a constant load on the unregulated voltage source?

A. A constant current source
B. A series regulator
C. A shunt current source
D. A shunt regulator

D   The shunt regulator regulates output voltage by providing a constant load to the unregulated (primary) voltage source. When more power is needed in the load, less power is diverted through the shunt element (usually a transistor). When less power is needed in the load, more power is diverted to the shunt element of the regulator.

E7D06 What is the purpose of Q1 in the circuit shown in Figure E7-3?

A. It provides negative feedback to improve regulation
B. It provides a constant load for the voltage source
C. It increases the current-handling capability of the regulator
D. It provides D1 with current

C   The Zener diode D1 will regulate voltage over a limited current range. Transistor Q1 increases that regulation range and thus increases the current-handling capability of the regulator circuit.

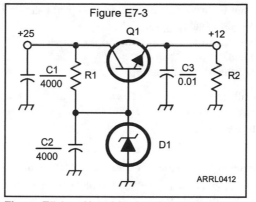

Figure E7-3 — Use this drawing to answer questions E7D06 through E7D13.

E7D07 What is the purpose of C2 in the circuit shown in Figure E7-3?
A. It bypasses hum around D1
B. It is a brute force filter for the output
C. To self-resonate at the hum frequency
D. To provide fixed DC bias for Q1

A The capacitor is used to filter out any ac signal components from the reference voltage.

E7D08 What type of circuit is shown in Figure E7-3?
A. Switching voltage regulator
B. Grounded emitter amplifier
C. Linear voltage regulator
D. Emitter follower

C Figure E7-3 is a linear voltage regulator.

E7D09 What is the purpose of C1 in the circuit shown in Figure E7-3?
A. It resonates at the ripple frequency
B. It provides fixed bias for Q1
C. It decouples the output
D. It filters the supply voltage

D The purpose of C1 in the circuit, a large capacitor of 4000 μF, is to filter the supply voltage from the power source.

E7D10   What is the purpose of C3 in the circuit shown in Figure E7-3?

A. It prevents self-oscillation
B. It provides brute force filtering of the output
C. It provides fixed bias for Q1
D. It clips the peaks of the ripple

A   C3, a small capacitor of 0.01 µF, shunts any high-frequency signals to ground and by so doing prevents self-oscillation should those signals be coupled to the base of Q1.

E7D11   What is the purpose of R1 in the circuit shown in Figure E7-3?

A. It provides a constant load to the voltage source
B. It couples hum to D1
C. It supplies current to D1
D. It bypasses hum around D1

C   R1 provides current to D1 to establish the regulator's reference voltage.

E7D12   What is the purpose of R2 in the circuit shown in Figure E7-3?

A. It provides fixed bias for Q1
B. It provides fixed bias for D1
C. It decouples hum from D1
D. It provides a constant minimum load for Q1

D   R2 is always connected, drawing power from Q1 at all times. This improves regulation by keeping the control circuitry active and thus able to respond better to sudden changes in the load.

E7D13   What is the purpose of D1 in the circuit shown in Figure E7-3?

A. To provide line voltage stabilization
B. To provide a voltage reference
C. Peak clipping
D. Hum filtering

B   D1 is a Zener diode and it provides a fixed reference voltage to the base of the transistor.

E7D14 What is one purpose of a "bleeder" resistor in a conventional (unregulated) power supply?

A. To cut down on waste heat generated by the power supply
B. To balance the low-voltage filament windings
C. To improve output voltage regulation
D. To boost the amount of output current

C Bleeder resistors provide a constant load to the supply, keeping the unregulated voltage within an acceptable range. Bleeder resistors also discharge the supply filter capacitors over a relatively long period. This is done for safety reasons.

E7D15 What is the purpose of a "step-start" circuit in a high-voltage power supply?

A. To provide a dual-voltage output for reduced power applications
B. To compensate for variations of the incoming line voltage
C. To allow for remote control of the power supply
D. To allow the filter capacitors to charge gradually

D Without a step-start or "soft-start" circuit in the supply, the discharged capacitors draw very high current at turn-on. This short-term current surge can be stressful to supply components, including the capacitors themselves. The step-start circuit reduces the surge by turning on the supply more gradually.

E7D16 When several electrolytic filter capacitors are connected in series to increase the operating voltage of a power supply filter circuit, why should resistors be connected across each capacitor?

A. To equalize, as much as possible, the voltage drop across each capacitor
B. To provide a safety bleeder to discharge the capacitors when the supply is off
C. To provide a minimum load current to reduce voltage excursions at light loads
D. All of these answers are correct

D Capacitors with the same nominal value are not identical in actual value. The resulting charge imbalance between several connected in series leads to an unequal charge distribution and voltage. Using a resistors in parallel with the series-connected capacitors also equalizes the voltage across the capacitors, ensuring that the voltage across them remains within their ratings. See also the discussion for E7D14.

E7D17 What is the primary reason that a high-frequency inverter type high-voltage power supply can be both less expensive and lighter in weight than a conventional power supply?

A. The inverter design does not require any output filtering
B. It uses a diode bridge rectifier for increased output
C. The high frequency inverter design uses much smaller transformers and filter components for an equivalent power output
D. It uses a large power-factor compensation capacitor to create "free" power from the unused portion of the AC cycle

C Because the cycles associated with the high-frequency inverter are much shorted than those of a 60-Hz linear supply, the transformers and capacitors can be much smaller and still perform equally well. The tradeoff is that the inverter circuit adds expense.

E7E Modulation and demodulation: reactance, phase and balanced modulators; detectors; mixer stages; DSP modulation and demodulation; software defined radio systems

E7E01 Which of the following can be used to generate FM-phone emissions?

A. A balanced modulator on the audio amplifier
B. A reactance modulator on the oscillator
C. A reactance modulator on the final amplifier
D. A balanced modulator on the oscillator

B A reactance modulator connected to the transmitter master oscillator's tank circuit is a simple and satisfactory device for producing true FM in an amateur transmitter.

E7E02 What is the function of a reactance modulator?

A. To produce PM signals by using an electrically variable resistance
B. To produce AM signals by using an electrically variable inductance or capacitance
C. To produce AM signals by using an electrically variable resistance
D. To produce PM signals by using an electrically variable inductance or capacitance

D A reactance modulator connected to the output of an oscillator circuit introduces a varying phase delay, creating a phase-modulated (PM) signal. As described in question E7E01, a reactance modulator can also produce FM.

E7E03   What is the fundamental principle of a phase modulator?

A. It varies the tuning of a microphone preamplifier to produce PM signals
B. It varies the tuning of an amplifier tank circuit to produce AM signals
C. It varies the tuning of an amplifier tank circuit to produce PM signals
D. It varies the tuning of a microphone preamplifier to produce AM signals

C   A phase modulator varies the tuning of an RF amplifier tank circuit to produce PM signals by introducing a variable phase shift in the signal path.

E7E04   What is one way a single-sideband phone signal can be generated?

A. By using a balanced modulator followed by a filter
B. By using a reactance modulator followed by a mixer
C. By using a loop modulator followed by a mixer
D. By driving a product detector with a DSB signal

A   You can generate a single-sideband phone signal by using a balanced modulator followed by a filter. The balanced modulator produces a double-sideband suppressed-carrier signal. The filter removes the unwanted sideband.

E7E05   What circuit is added to an FM transmitter to proportionally attenuate the lower audio frequencies?

A. A de-emphasis network
B. A heterodyne suppressor
C. An audio prescaler
D. A pre-emphasis network

D   A pre-emphasis circuit is used in a transmitter to attenuate lower frequencies so that the modulation is equalized.

E7E06   What circuit is added to an FM receiver to restore attenuated lower audio frequencies?

A. A de-emphasis network
B. A heterodyne suppressor
C. An audio prescaler
D. A pre-emphasis network

A   In a receiver, a de-emphasis circuit restores the audio frequency response to the recovered audio. You can remember the order from pre-emphasis applied before (or pre) transmission.

E7E07   What is one result of the process of mixing two signals?

    A. The elimination of noise in a wideband receiver by phase comparison

    B. The elimination of noise in a wideband receiver by phase differentiation

    C. The recovery of the intelligence from a modulated RF signal

    D. The creation of new signals at the sum and difference frequencies

D   In the mixing process, two signals are multiplied together. This produces mixing product signals at the sum and difference of the original two frequencies. In a practical mixer, output signals at the original two frequencies are also present, but at a reduced level.

E7E08   What are the principal frequencies that appear at the output of a mixer circuit?

    A. Two and four times the original frequency

    B. The sum, difference and square root of the input frequencies

    C. The original frequencies, and the sum and difference frequencies

    D. 1.414 and 0.707 times the input frequency

C   See the discussion for E7E07.

E7E09   What occurs when an excessive amount of signal energy reaches a mixer circuit?

    A. Spurious mixer products are generated

    B. Mixer blanking occurs

    C. Automatic limiting occurs

    D. A beat frequency is generated

A   Spurious mixer products will be produced if input-signal energy overloads the mixer circuit. The level of these spurious mixer products may increase to the point that they are detectable in the output. One result of these effects is that the receiver may suffer severe interference in the presence of extremely strong signals.

E7E10   What is the process of detection?

    A. The extraction of weak signals from noise

    B. The recovery of information from a modulated RF signal

    C. The modulation of a carrier

    D. The mixing of noise with a received signal

B   A detector is used to recover the information that has been superimposed through modulation on an RF signal.

E7E11   How does a diode detector function?

A. By rectification and filtering of RF signals
B. By breakdown of the Zener voltage
C. By mixing signals with noise in the transition region of the diode
D. By sensing the change of reactance in the diode with respect to frequency

A   A diode rectifies an AM signal, leaving the signal's envelope (containing the modulation information) and pulses at the AM carrier frequency. The envelope can be detected with a circuit that responds only to low frequencies, ignoring or filtering out the high-frequency pulses.

E7E12   Which of the following types of detector is well suited for demodulating SSB signals?

A. Discriminator
B. Phase detector
C. Product detector
D. Phase comparator

C   A product detector mixes an incoming signal with a local signal or beat-frequency oscillator (BFO). This converts the SSB signal from RF to audio frequencies.

E7E13   What is a frequency discriminator?

A. An FM generator circuit
B. A circuit for filtering two closely adjacent signals
C. An automatic band-switching circuit
D. A circuit for detecting FM signals

D   A frequency discriminator is found in an FM receiver where it functions as a detector, recovering the modulating information.

E7E14   Which of the following describes a common means of generating a SSB signal when using digital signal processing?

A. Mixing products are converted to voltages and subtracted by adder circuits
B. A frequency synthesizer removes the unwanted sidebands
C. Emulation of quartz crystal filter characteristics
D. The phasing or quadrature method

D   DSP circuits and software are well-suited to processing signals by varying their frequency and phase. This makes the phasing method a favorite for DSP-based equipment, but it is quite difficult for analog circuits.

E7E15   What is meant by "direct conversion" when referring to a software defined receiver?

A. Software is converted from source code to object code during operation of the receiver
B. Incoming RF is converted to the IF frequency by rectification to generate the control voltage for a voltage controlled oscillator
C. Incoming RF is mixed to "baseband" for analog-to-digital conversion and subsequent processing
D. Software is generated in machine language, avoiding the need for compilers

C   Instead of converting the signal to an intermediate frequency for filtering, a direct-conversion receiver converts it directly to audio or baseband. All filtering is then performed by DSP software before the information is output.

E7F   Frequency markers and counters: frequency divider circuits; frequency marker generators; frequency counters

E7F01   What is the purpose of a prescaler circuit?

A. It converts the output of a JK flip-flop to that of an RS flip-flop
B. It multiplies a higher frequency signal so a low-frequency counter can display the operating frequency
C. It prevents oscillation in a low-frequency counter circuit
D. It divides a higher frequency signal so a low-frequency counter can display the operating frequency

D   A prescaler is a frequency divider that allows a slower device to measure a high-frequency signal by "re-scaling" the signal's frequency, such as dividing it by 10.

E7F02   Which of the following would be used to reduce a signal's frequency by a factor of ten?

A. A preamp
B. A prescaler
C. A marker generator
D. A flip-flop

B   See the discussion for E7F01.

E7F03 What is the function of a decade counter digital IC?

A. It produces one output pulse for every ten input pulses
B. It decodes a decimal number for display on a seven-segment LED display
C. It produces ten output pulses for every input pulse
D. It adds two decimal numbers together

A A decade counter produces one output pulse for every 10 input pulses. Counter circuits are also called dividers.

E7F04 What additional circuitry must be added to a 100-kHz crystal-controlled marker generator so as to provide markers at 50 and 25 kHz?

A. An emitter-follower
B. Two frequency multipliers
C. Two flip-flops
D. A voltage divider

C You'll need to divide 100 kHz by both two and four to produce signals at 50 and 25 kHz. This will require two flip-flops. The first will divide the 100 kHz signal by two, producing markers at 50 kHz intervals. The second flip-flop also divides the 50 kHz signal by two, producing markers at 25 kHz intervals.

E7F05 Which of the following circuits can be combined to produce a 100 kHz fundamental signal with harmonics at 100 kHz intervals?

A. A 10 MHz oscillator and a flip-flop
B. A 1 MHz oscillator and a decade counter
C. A 1 MHz oscillator and a flip-flop
D. A 100 kHz oscillator and a phase detector

B If the 1 MHz signal is divided by 10 then the output signal will be 100 kHz. This output signal is a digital signal with sharp rising and falling edges, so it will produce harmonics every 100 kHz.

E7F06 Which of these choices best describes a crystal marker generator?

A. A low-stability oscillator that sweeps through a band of frequencies
B. An oscillator often used in aircraft to determine the craft's location relative to the inner and outer markers at airports
C. A crystal-controlled oscillator with an output frequency and amplitude that can be varied over a wide range
D. A crystal-controlled oscillator that generates a series of reference signals at known frequency intervals

D A crystal-controlled marker generator is a high-stability oscillator that generates a series of reference signals at known frequency intervals.

E7F07 Which type of circuit would be a good choice for generating a series of harmonically related receiver calibration signals?

A. A Wein-bridge oscillator followed by a class-A amplifier
B. A Foster-Seeley discriminator
C. A phase-shift oscillator
D. A crystal oscillator followed by a frequency divider

D A sinusoidal crystal oscillator generates a single frequency (no harmonics), so it is not a good marker generator by itself. However, by connecting the output of the oscillator to a digital frequency divider circuit a signal containing many harmonics is produced — excellent for use as markers.

E7F08 What is one purpose of a marker generator?

A. To add audio markers to an oscilloscope
B. To provide a frequency reference for a phase locked loop
C. To provide a means of calibrating a receiver's frequency settings
D. To add time signals to a transmitted signal

C A marker generator is used to calibrate receiver frequency settings because it produces signals at known frequencies.

E7F09 What determines the accuracy of a frequency counter?

A. The accuracy of the time base
B. The speed of the logic devices used
C. Accuracy of the AC input frequency to the power supply
D. Proper balancing of the mixer diodes

A The accuracy of a frequency counter depends on its internal crystal reference oscillator (time base) which establishes all timing for the counter circuits.

E7F10 How does a conventional frequency counter determine the frequency of a signal?

A. It counts the total number of pulses in a circuit
B. It monitors a WWV reference signal for comparison with the measured signal
C. It counts the number of input pulses occurring within a specific period of time
D. It converts the phase of the measured signal to a voltage which is proportional to the frequency

C By knowing the amount of time pulses are counted (the gate period), it is easy to convert a pulse total to frequency.

E7F11 What is the purpose of a frequency counter?

A. To provide a digital representation of the frequency of a signal
B. To generate a series of reference signals at known frequency intervals
C. To display all frequency components of a transmitted signal
D. To provide a signal source at a very accurate frequency

A The purpose of a frequency counter is to indicate the frequency of its strongest input signal.

E7F12 What alternate method of determining frequency, other than by directly counting input pulses, is used by some frequency counters?

A. GPS averaging
B. Period measurement
C. Prescaling
D. D/A conversion

B For very slow signals, instead of counting many pulses and then averaging over the gate period, it is more convenient and often just as accurate to measure the period of the signal and then invert it to get frequency.

E7F13 What is an advantage of a period-measuring frequency counter over a direct-count type?

A. It can run on battery power for remote measurements
B. It does not require an expensive high-precision time base
C. It provides improved resolution of signals within a comparable time period
D. It can directly measure the modulation index of an FM transmitter

C See the discussion for E7F12.

E7G Active filters and op-amps: active audio filters; characteristics; basic circuit design; operational amplifiers

E7G01   What determines the gain and frequency characteristics of an op-amp RC active filter?

A. The values of capacitors and resistors built into the op-amp
B. The values of capacitors and resistors external to the op-amp
C. The input voltage and frequency of the op-amp's DC power supply
D. The output voltage and smoothness of the op-amp's DC power supply

B   The gain and frequency characteristics of an op-amp RC active filter are determined by values of capacitances and resistances external to the op-amp.

E7G02   What causes ringing in a filter?

A. The slew rate of the filter
B. The bandwidth of the filter
C. The frequency and phase response of the filter
D. The gain of the filter

C   Filter ringing occurs when the Q of a filter is too high, leading to a very narrow bandwidth, or creating peaks in the filter's amplitude response over frequency.

E7G03   What are the advantages of using an op-amp instead of LC elements in an audio filter?

A. Op-amps are more rugged and can withstand more abuse than can LC elements
B. Op-amps are fixed at one frequency
C. Op-amps are available in more varieties than are LC elements
D. Op-amps exhibit gain rather than insertion loss

D   One of the advantages of using an op-amp instead of LC elements in an audio filter is that op-amp filters can have gain, amplifying the signal at the same time as filtering it.

E7G04 Which of the following capacitor types is best suited for use in high-stability op-amp RC active filter circuits?

A. Electrolytic
B. Disc ceramic
C. Polystyrene
D. Paper dielectric

C Capacitors that you use in a high-stability op-amp RC active filter circuit should be high Q and temperature stable. Polystyrene units are excellent for such use.

E7G05 How can unwanted ringing and audio instability be prevented in a multi-section op-amp RC audio filter circuit?

A. Restrict both gain and Q
B. Restrict gain, but increase Q
C. Restrict Q, but increase gain
D. Increase both gain and Q

A To avoid unwanted ringing and audio instability in a multisection op-amp RC audio filter circuit, you should limit (restrict) both the gain and the Q of the circuit.

E7G06 What steps are typically followed when selecting the external components for an op-amp RC active filter?

A. Standard capacitor values are chosen first, the resistances are calculated, and resistors of the nearest standard value are used
B. Standard resistor values are chosen first, the capacitances are calculated, and capacitors of the nearest standard value are used
C. Standard resistor and capacitor values are used, the circuit is tested, and additional resistors are added to make any needed adjustments
D. Standard resistor and capacitor values are used, the circuit is tested, and additional capacitors are added to make any needed adjustments

A Standard capacitor values are chosen first because capacitors are not available in as many values as resistors. Then calculate resistances and use the nearest standard value.

E7G07 Which of the following is the most appropriate use of an op-amp RC active filter?

A. As a high-pass filter used to block RFI at the input to receivers
B. As a low-pass filter used between a transmitter and a transmission line
C. For smoothing power-supply output
D. As an audio receiving filter

D Op-amp RC active filters are well suited for audio-frequency applications.

E7G08 Which of the following is a type of active op-amp filter circuit?

A. Regenerative feedback resonator
B. Helical resonator
C. Gilbert cell
D. Sallen-Key

D Sallen-Key filter equations for active op-amp filter circuits are named after the designers who first derived them. Other sets of equations and circuits can also create band-pass filters, but the Sallen-Key equations are the easiest to use.

E7G09 What voltage gain can be expected from the circuit in Figure E7-4 when R1 is 10 ohms and RF is 470 ohms?

A. 0.21
B. 94
C. 47
D. 24

C The voltage gain for the circuit in the drawing is

$$V_{GAIN} = \frac{-R_F}{R1} = \frac{-470}{10} = -47$$

The minus sign indicates that this is an inverting amplifier. That means that a positive input gives a negative output and vice versa. In stating gain, it is usual to ignore the inversion, and following that practice the gain is 47.

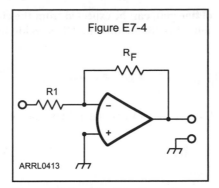

Figure E7-4

Figure E7-4 — Use this figure to answer questions E7G09 and E7G11 through E7D13.

E7G10 How does the gain of a theoretically ideal operational amplifier vary with frequency?

A. It increases linearly with increasing frequency
B. It decreases linearly with increasing frequency
C. It decreases logarithmically with increasing frequency
D. It does not vary with frequency

D In an *ideal* operational amplifier, there is no change in gain as the frequency changes. In other words, the frequency response is flat.

E7G11 What will be the output voltage of the circuit shown in Figure E7-4 if R1 is 1000 ohms, RF is 10,000 ohms, and 0.23 volts is applied to the input?

A. 0.23 volts
B. 2.3 volts
C. −0.23 volts
D. −2.3 volts

D Use the voltage gain equation from the earlier question. The output voltage is equal to voltage gain times the input voltage. So,

$$V_{OUT} = \frac{-R_F}{R1} \times V_{IN}$$

Using the values given,

$$V_{OUT} = \frac{-10,000}{1000} \times 0.23 = -2.3 \text{ V}$$

Practical CIrcuits   **35**

E7G12 What voltage gain can be expected from the circuit in Figure E7-4 when R1 is 1800 ohms and RF is 68 kilohms?

A. 1
B. 0.03
C. 38
D. 76

C Use the same formula as before. Using the resistor values given, you'll find

$$V_{GAIN} = \frac{-R_F}{R1} = \frac{-68,000}{1800} = -38$$

As before, you can ignore the minus sign when stating the circuit gain.

E7G13 What voltage gain can be expected from the circuit in Figure E7-4 when R1 is 3300 ohms and RF is 47 kilohms?

A. 28
B. 14
C. 7
D. 0.07

B Another question using the same formula. Using the resistor values given, you'll find

$$V_{GAIN} = \frac{-R_F}{R1} = \frac{-47,000}{3300} = -14$$

Ignore the minus sign as you did before.

E7G14 What is an operational amplifier?

A. A high-gain, direct-coupled differential amplifier whose characteristics are determined by components external to the amplifier
B. A high-gain, direct-coupled audio amplifier whose characteristics are determined by components external to the amplifier
C. An amplifier used to increase the average output of frequency modulated amateur signals to the legal limit
D. A program subroutine that calculates the gain of an RF amplifier

A The operational amplifier (op amp) is a high-gain, direct-coupled, differential amplifier that will amplify dc signals as well as ac signals. The key item to spot here is "differential amplifier."

E7G15 What is meant by the term "op-amp input-offset voltage"?

A. The output voltage of the op-amp minus its input voltage
B. The difference between the output voltage of the op-amp and the input voltage required in the immediately following stage
C. The potential between the amplifier input terminals of the op-amp in a closed-loop condition
D. The potential between the amplifier input terminals of the op-amp in an open-loop condition

C Op-amp input-offset voltage is the voltage between the amplifier input terminals of the op-amp in a closed-loop condition. Offset results from imbalance between the differential input transistors in the IC.

E7G16 What is the typical input impedance of an integrated circuit op-amp?

A. 100 ohms
B. 1000 ohms
C. Very low
D. Very high

D The theoretical (perfect) op-amp has infinite input impedance. A good approximation to this is "very high."

E7G17 What is the typical output impedance of an integrated circuit op-amp?

A. Very low
B. Very high
C. 100 ohms
D. 1000 ohms

A The output is the opposite of the input, and the ideal op amp has a very low output impedance.

E7H Oscillators and signal sources: types of oscillators; synthesizers and phase-locked loops; direct digital synthesizers

E7H01   What are three major oscillator circuits often used in Amateur Radio equipment?

A. Taft, Pierce and negative feedback
B. Pierce, Fenner and Beane
C. Taft, Hartley and Pierce
D. Colpitts, Hartley and Pierce

D   Colpitts, Hartley and Pierce are the most popular oscillator circuits, named after their inventors.

E7H02   What condition must exist for a circuit to oscillate?

A. It must have at least two stages
B. It must be neutralized
C. It must have a positive feedback loop with a gain greater than 1
D. It must have negative feedback sufficient to cancel the input signal

C   An oscillator requires positive feedback, which reinforces oscillation. You'll need enough gain to overcome feedback loop losses in order to sustain oscillation.

E7H03   How is positive feedback supplied in a Hartley oscillator?

A. Through a tapped coil
B. Through a capacitive divider
C. Through link coupling
D. Through a neutralizing capacitor

A   Hartley starts with H, and so does Henry — the basic unit of inductance. You can use that as a mnemonic to help you remember that the Hartley oscillator uses a tapped coil to provide the positive feedback needed for operation.

E7H04   How is positive feedback supplied in a Colpitts oscillator?

A. Through a tapped coil
B. Through link coupling
C. Through a capacitive divider
D. Through a neutralizing capacitor

C   Colpitts starts with C, and so does capacitor. You can use that as a mnemonic to help you remember that the Colpitts oscillator uses a capacitive voltage divider to provide the positive feedback needed for operation.

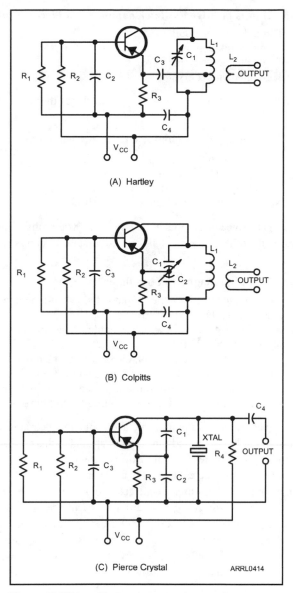

(A) Hartley

(B) Colpitts

(C) Pierce Crystal

ARRL0414

Figure E7H01 — Three common types of transistor oscillator circuits are the Hartley (A), Colpitts (B), and Pierce (C).

E7H05 How is positive feedback supplied in a Pierce oscillator?

A. Through a tapped coil
B. Through link coupling
C. Through a neutralizing capacitor
D. Through a quartz crystal

D The Pierce oscillator provides positive feedback through a quartz crystal.

E7H06 Which type of oscillator circuits are commonly used in VFOs?

A. Pierce and Zener
B. Colpitts and Hartley
C. Armstrong and deForest
D. Negative feedback and Balanced feedback

B The right choice is Colpitts and Hartley, but you probably guessed that, recognizing the other answers as all or partly not oscillators.

E7H07 What is a magnetron oscillator?

A. An oscillator in which the output is fed back to the input by the magnetic field of a transformer
B. An crystal oscillator in which variable frequency is obtained by placing the crystal in a strong magnetic field
C. A UHF or microwave oscillator consisting of a diode vacuum tube with a specially shaped anode, surrounded by an external magnet
D. A reference standard oscillator in which the oscillations are synchronized by magnetic coupling to a rubidium gas tube

C A magnetron tube is a diode. The anode of a magnetron is not a flat plate as in most tubes, but is formed into a set of specially-shaped resonant cavities that surround the cathode. A strong external magnet creates a magnetic field inside the tube. When the cathode is heated and raised to a high negative voltage, it emits electrons that are attracted to the anode. The strong magnetic field causes the electrons to revolve around the cathode as they travel toward the anode. As they spiral outward, the electrons emit RF energy, creating an oscillating electric field. The resonant cavities act as the filter for the oscillator, reinforcing a single frequency of oscillation. RF power is extracted from the tube through a glass window through which the RF energy is coupled into a waveguide or an oven.

E7H08   What is a Gunn diode oscillator?

A. An oscillator based on the negative resistance properties of properly-doped semiconductors
B. An oscillator based on the argon gas diode
C. A highly stable reference oscillator based on the tee-notch principle
D. A highly stable reference oscillator based on the hot-carrier effect

A   A Gunn diode oscillator makes use of a resonant cavity to control the frequency of the generated signal. It does so by placing an amplifier inside the cavity. Gunn diodes (named for their inventor) are semiconductor diodes with special types of doping. As forward bias voltage is increased across these diodes, a point is reached at which current decreases with increasing voltage. This is negative resistance. If such a device is placed inside a tuned cavity and forward biased, the negative resistance acts as an amplifier, with the cavity providing the necessary resonant circuit. The cavity acts both as a feedback circuit to the diode and as a resonant circuit to control the frequency. If the negative resistance is strong enough, the combination creates an oscillator.

E7H09   What type of frequency synthesizer circuit uses a stable voltage-controlled oscillator, programmable divider, phase detector, loop filter and a reference frequency source?

A. A direct digital synthesizer
B. A hybrid synthesizer
C. A phase locked loop synthesizer
D. A diode-switching matrix synthesizer

C   A phase-locked loop synthesizer uses a stable voltage-controlled oscillator, programmable divider, phase detector, loop filter and a reference frequency source. See Figure E7H09.

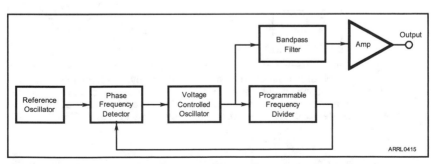

Figure E7H09 — Block diagram of a phase-locked loop frequency synthesizer.

E7H10 What type of frequency synthesizer circuit uses a phase accumulator, lookup table, digital to analog converter and a low-pass anti-alias filter?

A. A direct digital synthesizer
B. A hybrid synthesizer
C. A phase locked loop synthesizer
D. A diode-switching matrix synthesizer

A   A direct digital synthesizer uses a phase accumulator, lookup table, digital to analog converter and a low-pass antialias filter. See Figure E7H10.

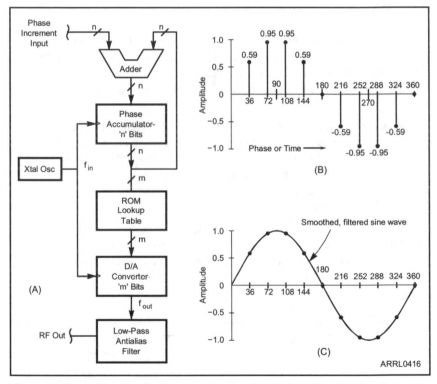

Figure E7H10 — The block diagram of a direct digital synthesizer is shown at (A). At (B), the amplitude values found in the ROM lookup table for a particular sine wave being generated by the synthesizer. The smoothed output signal from the synthesizer after it goes through the low-pass antialias filter is shown at (C).

E7H11 What information is contained in the lookup table of a direct digital frequency synthesizer?

A. The phase relationship between a reference oscillator and the output waveform
B. The amplitude values that represent a sine-wave output
C. The phase relationship between a voltage-controlled oscillator and the output waveform
D. The synthesizer frequency limits and frequency values stored in the radio memories

B   The lookup table of a direct digital frequency synthesizer is a list of sine-wave values at various phase angles.

E7H12 What are the major spectral impurity components of direct digital synthesizers?

A. Broadband noise
B. Digital conversion noise
C. Spurs at discrete frequencies
D. Nyquist limit noise

C   The major spectral impurity components are spurious signals, or spurs, at specific discrete frequencies related to the frequencies of the digital circuits that make up the synthesizer.

E7H13 Which of these circuits would be classified as a principal component of a direct digital synthesizer (DDS)?

A. Phase splitter
B. Hex inverter
C. Chroma demodulator
D. Phase accumulator

D   The phase accumulator controls the sine wave value selected from the lookup table.

E7H14 What circuit is often used in conjunction with a direct digital synthesizer (DDS) to expand the available tuning range?

A. Binary expander
B. J-K flip-flop
C. Phase locked loop
D. Compander

C   A DDS synthesizer can be substituted for either of the oscillators in a phase-locked loop (PLL). This greatly increases the frequency range over which the PLL can operate.

E7H15 What is the capture range of a phase-locked loop circuit?

    A. The frequency range over which the circuit can lock
    B. The voltage range over which the circuit can lock
    C. The input impedance range over which the circuit can lock
    D. The range of time it takes the circuit to lock

A A PLL is designed to operate over a certain frequency range. The capture range is the region in the frequency domain over which the circuit maintains control (lock) of its voltage-controlled oscillator.

E7H16 What is a phase-locked loop circuit?

    A. An electronic servo loop consisting of a ratio detector, reactance modulator, and voltage-controlled oscillator
    B. An electronic circuit also known as a monostable multivibrator
    C. An electronic servo loop consisting of a phase detector, a low-pass filter and voltage-controlled oscillator
    D. An electronic circuit consisting of a precision push-pull amplifier with a differential input

C The phase-locked loop is an electronic servo loop consisting of a phase detector, low-pass filter and voltage-controlled oscillator. See the discussion for E7H09.

E7H17 Which of these functions can be performed by a phase-locked loop?

    A. Wide-band AF and RF power amplification
    B. Comparison of two digital input signals, digital pulse counter
    C. Photovoltaic conversion, optical coupling
    D. Frequency synthesis, FM demodulation

D Phase-locked loops are used in frequency synthesis and FM demodulation applications. In the first example, variable frequency dividers are used so that the voltage-controlled oscillator operates at a selectable frequency. In the second, the loop's error signal contains the modulating signal as the PLL tracks the FM signal frequency.

E7H18 Why is a stable reference oscillator normally used as part of a phase locked loop (PLL) frequency synthesizer?

A. Any amplitude variations in the reference oscillator signal will prevent the loop from locking to the desired signal
B. Any phase variations in the reference oscillator signal will produce phase noise in the synthesizer output
C. Any phase variations in the reference oscillator signal will produce harmonic distortion in the modulating signal
D. Any amplitude variations in the reference oscillator signal will prevent the loop from changing frequency

B Phase noise in a synthesizer output used to drive a transmitter can cause interference to nearby signals. Phase noise in an oscillator used for receiver mixing causes nearby signals to interfere with reception of the desired signal.

E7H19 Why is a phase-locked loop often used as part of a variable frequency synthesizer for receivers and transmitters?

A. It generates FM sidebands
B. It eliminates the need for a voltage controlled oscillator
C. It makes it possible for a VFO to have the same degree of stability as a crystal oscillator
D. It can be used to generate or demodulate SSB signals by quadrature phase synchronization

C The PLL is locked to a crystal-controlled reference oscillator, so it has the advantage of both variable frequency operation and crystal-controlled stability.

E7H20 What are the major spectral impurity components of phase-locked loop synthesizers?

A. Broadband noise
B. Digital conversion noise
C. Spurs at discrete frequencies
D. Nyquist limit noise

A A phase-locked loop synthesizer is constantly "correcting" the output signal frequency. This results in slight phase shifts from one cycle to the next. This in turn results in phase noise, which is a broadband noise around the desired output frequency and which is the major spectral impurity component of a PLL synthesizer.

Signals and Emissions

There will be four questions on your Extra class examination from the Signals and Emissions subelement. These four questions will be taken from the four groups of questions labeled E8A through E8D.

E8A AC waveforms: sine, square, sawtooth and irregular waveforms; AC measurements; average and PEP of RF signals; pulse and digital signal waveforms

E8A01 What type of wave is made up of a sine wave plus all of its odd harmonics?
A. A square wave
B. A sine wave
C. A cosine wave
D. A tangent wave

A A square wave is made up of a fundamental frequency and all its odd harmonics.

E8A02 What type of wave has a rise time significantly faster than its fall time (or vice versa)?
A. A cosine wave
B. A square wave
C. A sawtooth wave
D. A sine wave

C A sawtooth wave appears on an oscilloscope as a straight-line waveform that has a rise time faster than the fall time (or vice-versa). You can see a sawtooth wave in Figure E8A02.

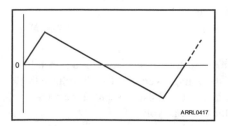

Figure E8A02 — Sawtooth waves are straight-line waveforms with unequal rise and fall times. Sawtooth waves consist of both odd and even harmonics, as well as the fundamental.

E8A03 What type of wave is made up of sine waves of a given fundamental frequency plus all its harmonics?

A. A sawtooth wave
B. A square wave
C. A sine wave
D. A cosine wave

A A sawtooth wave is made up of sine waves of a fundamental frequency and all harmonics. Look at Figure E8A03 to see how the waveform will look on a spectrum analyzer. You see components at 10 Hz, 20 Hz, 30 Hz, 40 Hz, 50 Hz, 60 Hz, 70 Hz, and on out to infinity. Since these are both the even and the odd multiples of the fundamental frequency (10 Hz), the sawtooth wave is said to be made up of the fundamental frequency and all harmonics.

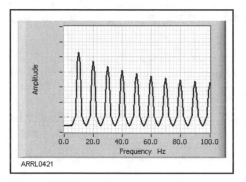

Figure E8A03 — Spectrum analyzer display of a 10-Hz sawtooth wave. The sawtooth waveform's spectrum consists of components that include the fundamental at 10 Hz plus all of the harmonics at 20 Hz, 30 Hz, 40 Hz, and so on.

E8A04 What is the equivalent to the root-mean-square value of an AC voltage?

A. The AC voltage found by taking the square of the average value of the peak AC voltage
B. The DC voltage causing the same amount of heating in a given resistor as the corresponding peak AC voltage
C. The DC voltage causing the same amount of heating in a resistor as the corresponding RMS AC voltage
D. The AC voltage found by taking the square root of the average AC value

C When an ac voltage is applied to a resistor, the resistor will dissipate energy in the form of heat, just as if the voltage were dc. The dc voltage that would cause identical heating in the ac-excited resistor is called the root-mean-square (RMS) or effective value of the ac voltage.

E8A05 What would be the most accurate way of measuring the RMS voltage of a complex waveform?

A. By using a grid dip meter
B. By measuring the voltage with a D'Arsonval meter
C. By using an absorption wavemeter
D. By measuring the heating effect in a known resistor

D   This is a variation of the previous question. You should have no trouble identifying the heating effect in a known resistor as the most accurate way of measuring the RMS voltage of a complex waveform.

E8A06 What is the approximate ratio of PEP-to-average power in a typical voice-modulated single-sideband phone signal?

A. 2.5 to 1
B. 25 to 1
C. 1 to 1
D. 100 to 1

A   PEP (peak envelope power) is the average power during one cycle at a modulation peak. These envelope peaks occur sporadically during voice transmission. In Figure E8A06, you can see a couple of examples of the relationship of PEP to average power in SSB signals. The ratio of peak-to-average amplitude varies widely with voices of different characteristics. The PEP of an SSB signal may be about 2 to 3 times greater than the average power output.

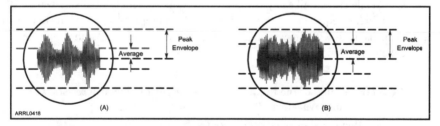

Figure E8A06 — Two RF envelope patterns that show the difference between average and peak levels. In each case, the RF amplitude is plotted as a function of time, as on an oscilloscope display. In (B), the average level is greater, which raises the average output power compared to the peak value.

E8A07 What determines the PEP-to-average power ratio of a single-sideband phone signal?

A. The frequency of the modulating signal
B. The characteristics of the modulating signal
C. The degree of carrier suppression
D. The amplifier gain

B The PEP-to-average power ratio is determined by the shape of the voice waveform. In other words, the ratio is determined by the modulating speech characteristics.

E8A08 What is the period of a wave?

A. The time required to complete one cycle
B. The number of degrees in one cycle
C. The number of zero crossings in one cycle
D. The amplitude of the wave

A The period of a wave is the time required to complete one cycle of that wave (regardless of its shape).

E8A09 What type of waveform is produced by human speech?

A. Sinusoidal
B. Logarithmic
C. Irregular
D. Trapezoidal

C While speech, like any complex waveform, is composed of sine waves of different frequencies, the combinations do not repeat like a waveform that is unchanging with time, such as a square wave. Waveforms of this type are called irregular.

E8A10 Which of the following is a distinguishing characteristic of a pulse waveform?

A. Regular sinusoidal oscillations
B. Narrow bursts of energy separated by periods of no signal
C. A series of tones that vary between two frequencies
D. A signal that contains three or more discrete tones

B The definition of a pulse is a short burst of energy, so a pulse waveform is one composed of short periods of signal separated by longer periods of no signal.

E8A11 What is one use for a pulse modulated signal?

A. Linear amplification
B. PSK31 data transmission
C. Multiphase power transmission
D. Digital data transmission

D Signals modulated by pulse waveforms are well-suited to sending digital data, since the data itself consists of on/off patterns.

E8A12 What type of information can be conveyed using digital waveforms?

A. Human speech
B. Video signals
C. Data
D. All of these answers are correct

D Any type of analog signal can be converted into digital information that can then be transmitted as data in a digital waveform.

E8A13 What is an advantage of using digital signals instead of analog signals to convey the same information?

A. Less complex circuitry is required for digital signal generation and detection
B. Digital signals always occupy a narrower bandwidth
C. Digital signals can be regenerated multiple times without error
D. All of these answers are correct

C It is actually more complicated to send and receive signals containing digital data than it is for analog modulation techniques. Nevertheless, the benefits of being able to replicate digital signals exactly for retransmission or storage any number of times outweighs the disadvantages of the additional complexity.

E8A14 Which of these methods is commonly used to convert analog signals to digital signals?

A. Sequential sampling
B. Harmonic regeneration
C. Level shifting
D. Phase reversal

A The process of generating a sequence of values that represent periodic measurements of a continuous analog waveform is called "sequential sampling." Each value in the sequence is a single measurement of the instantaneous amplitude of the waveform at a sampling time. When we make the measurements continually at regular intervals, the result is a sequence of values representing the amplitude of the signal at evenly spaced times.

E8A15 What would the waveform of a digital data stream signal look like on a conventional oscilloscope?

A. A series of sine waves with evenly spaced gaps
B. A series of pulses with varying patterns
C. A running display of alpha-numeric characters
D. None of the above; this type of signal cannot be seen on a conventional oscilloscope

B Digital data is composed of varying patterns of on/off signals, most commonly represented as high or low voltages. An oscilloscope would display digital data as varying patterns of voltage pulses.

E8B Modulation and demodulation: modulation methods; modulation index and deviation ratio; pulse modulation; frequency and time division multiplexing

E8B01 What is the term for the ratio between the frequency deviation of an RF carrier wave, and the modulating frequency of its corresponding FM-phone signal?

A. FM compressibility
B. Quieting index
C. Percentage of modulation
D. Modulation index

D Modulation index is represented by the symbol β and is calculated as $\beta = \Delta f/f_m$. That formula tells you that modulation index is the ratio between the deviation of the frequency-modulated signal (Δf) and the modulating frequency (f_m).

E8B02 How does the modulation index of a phase-modulated emission vary with RF carrier frequency (the modulated frequency)?

A. It increases as the RF carrier frequency increases
B. It decreases as the RF carrier frequency increases
C. It varies with the square root of the RF carrier frequency
D. It does not depend on the RF carrier frequency

D Reviewing the formula for modulation index in the previous question, you can see that modulation index does not depend on the RF carrier frequency.

E8B03 What is the modulation index of an FM-phone signal having
a maximum frequency deviation of 3000 Hz either side of the
carrier frequency, when the modulating frequency is 1000 Hz?

A. 3
B. 0.3
C. 3000
D. 1000

A For FM systems being modulated by a tone, the modulation index is
given by

$$\beta = \frac{\Delta f}{f_m}$$

where Δf = peak deviation in hertz and f_m = modulating frequency in hertz at
any given instant.

Using the values given in the question gives

$$\beta = \frac{3000 \text{ Hz}}{1000 \text{ Hz}} = 3$$

E8B04 What is the modulation index of an FM-phone signal having
a maximum carrier deviation of plus or minus 6 kHz when
modulated with a 2-kHz modulating frequency?

A. 6000
B. 3
C. 2000
D. 1/3

B Use the same formula that you did for the previous question. The
modulation index is given by

$$\beta = \frac{\Delta f}{f_m}$$

Using the parameters from the question gives

$$\beta = \frac{6000 \text{ Hz}}{2000 \text{ Hz}} = 3$$

E8B05   What is the deviation ratio of an FM-phone signal having
a maximum frequency swing of plus-or-minus 5 kHz and
accepting a maximum modulation rate of 3 kHz?

A. 60
B. 0.167
C. 0.6
D. 1.67

D   Deviation ratio is described by the formula

$$D = \frac{\Delta f}{f_{MAX}}$$

where Δf = peak deviation in hertz and f_{MAX} = maximum modulating
frequency in hertz.

It measures the ratio of the maximum carrier frequency deviation to the
highest audio modulating frequency. The formula for deviation ratio is
similar to the one for the modulation index. However, you have to replace the
instantaneous modulating frequency, f_m, with the maximum frequency in the
modulating signal.

Using the values from the question gives

$$D = \frac{5000 \text{ Hz}}{3000 \text{ Hz}} = 1.67$$

E8B06   What is the deviation ratio of an FM-phone signal having a
maximum frequency swing of plus or minus 7.5 kHz and
accepting a maximum modulation frequency of 3.5 kHz?

A. 2.14
B. 0.214
C. 0.47
D. 47

A   Use the formula given in the previous question to compute the deviation
ratio

$$D = \frac{7.5 \text{ kHz}}{3.5 \text{ kHz}} = 2.1$$

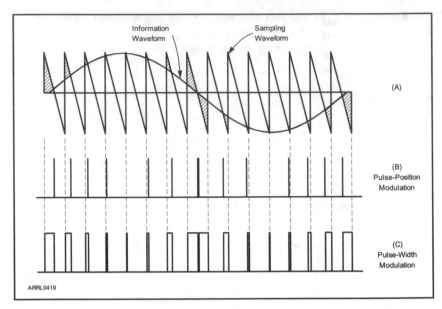

Figure E8B07 — An information signal and a sampling waveform are shown at (A). (B) shows how the sampled information signal can produce pulse-position modulation. (C) shows how sampling can produce pulse-width modulation.

E8B07 When using a pulse-width modulation system, why is the transmitter's peak power greater than its average power?

A. The signal duty cycle is less than 100%

B. The signal reaches peak amplitude only when voice modulated

C. The signal reaches peak amplitude only when voltage spikes are generated within the modulator

D. The signal reaches peak amplitude only when the pulses are also amplitude modulated

A In a pulse width-modulation system the width (duration) of pulses varies with the applied modulation. The transmitted pulses are all of the same amplitude. In most cases, the duty cycle of the transmission is very low. A pulse of relatively short duration is transmitted, with a relatively long period of time separating each pulse. This causes the peak power of a pulse-modulated signal to be much greater than its average power. See Figure E8B07.

E8B08 What parameter does the modulating signal vary in a pulse-position modulation system?

A. The number of pulses per second
B. The amplitude of the pulses
C. The duration of the pulses
D. The time at which each pulse occurs

D Pulse-position modulation varies the pulse position (or the time at which each pulse occurs) according to the characteristic of the modulating signal.

E8B09 How are the pulses of a pulse-modulated signal usually transmitted?

A. A pulse of relatively short duration is sent; a relatively long period of time separates each pulse
B. A pulse of relatively long duration is sent; a relatively short period of time separates each pulse
C. A group of short pulses are sent in a relatively short period of time; a relatively long period of time separates each group
D. A group of short pulses are sent in a relatively long period of time; a relatively short period of time separates each group

A A typical pulse modulated system transmits pulses of relatively short duration. A relatively long period of time separates each pulse.

E8B10 What is meant by deviation ratio?

A. The ratio of the audio modulating frequency to the center carrier frequency
B. The ratio of the maximum carrier frequency deviation to the highest audio modulating frequency
C. The ratio of the carrier center frequency to the audio modulating frequency
D. The ratio of the highest audio modulating frequency to the average audio modulating frequency

B See the discussion for E8B05.

E8B11 Which of these methods can be used to combine several separate analog information streams into a single analog radio frequency signal?

A. Frequency shift keying
B. A diversity combiner
C. Frequency division multiplexing
D. Pulse compression

C Multiplexing is defined as combining several streams of information into a single stream. Of the various schemes for combining the streams, frequency division multiplexing (FDM) is often used to combine analog information streams for analog modulation. The usual method is to shift the information streams to different frequencies, add them together to create a "baseband" signal, then modulate a carrier with the baseband signal.

E8B12 Which of the following describes frequency division multiplexing?

A. The transmitted signal jumps from band to band at a predetermined rate
B. Two or more information streams are merged into a "baseband", which then modulates the transmitter
C. The transmitted signal is divided into packets of information
D. Two or more information streams are merged into a digital combiner, which then pulse position modulates the transmitter

B See the discussion for E8B11.

E8B13 What is time division multiplexing?

A. Two or more data streams are assigned to discrete sub-carriers on an FM transmitter
B. Two or more signals are arranged to share discrete time slots of a digital data transmission
C. Two or more data streams share the same channel by transmitting time of transmission as the sub-carrier
D. Two or more signals are quadrature modulated to increase bandwidth efficiency

B In time division multiplexing (TDM), the information streams are divided into "time slices." The individual slices are then interleaved into a single information stream. This process works best for digital information.

E8C   Digital signals: digital communications modes; CW; information rate vs. bandwidth; spread-spectrum communications; modulation methods

E8C01   Which one of the following digital codes consists of elements having unequal length?

A. ASCII
B. AX.25
C. Baudot
D. Morse code

D   Morse code dits and dahs are of different (unequal) length. ASCII, AX.25 (packet) and Baudot code elements are of uniform length.

E8C02   What are some of the differences between the Baudot digital code and ASCII?

A. Baudot uses four data bits per character, ASCII uses seven; Baudot uses one character as a shift code, ASCII has no shift code
B. Baudot uses five data bits per character, ASCII uses seven; Baudot uses two characters as shift codes, ASCII has no shift code
C. Baudot uses six data bits per character, ASCII uses seven; Baudot has no shift code, ASCII uses two characters as shift codes
D. Baudot uses seven data bits per character, ASCII uses eight; Baudot has no shift code, ASCII uses two characters as shift codes

B   It will help if you can remember Baudot 5 and ASCII 7. The 5 data bits in the Baudot (ITA2) code can be arranged into 32 different combinations. (That's not enough to cover all the letters and the digits 0 through 9.) Fortunately, some of these combinations can be used twice, that's done by using the letters and figures shift characters to shift between those combinations. With 7 data bits, ASCII code has 128 distinctive patterns, and for that reason doesn't need shift codes.

E8C03   What is one advantage of using the ASCII code for data communications?

A. It includes built-in error-correction features
B. It contains fewer information bits per character than any other code
C. It is possible to transmit both upper and lower case text
D. It uses one character as a shift code to send numeric and special characters

C   With 7 data bits, ASCII code has 128 distinctive patterns. That makes it possible to transmit both upper and lower case text and still have patterns left for numbers, punctuation and special purpose characters.

E8C04 This question has been removed by the Question Pool Committee.

E8C05 What technique is used to minimize the bandwidth requirements of a PSK-31 signal?

A. Zero-sum character encoding
B. Reed-Solomon character encoding
C. Use of sinusoidal data pulses
D. Use of trapezoidal data pulses

C If a modulating signal has sharp edges, such as a for a digital data stream, then the resulting modulated signal must have enough bandwidth to reproduce the high-frequency components of the modulating signal that create the sharp edges. By using a more rounded modulating signal, less bandwidth is required for the modulated signal. PSK31 uses sinusoidally shaped pulses as its method of minimizing bandwidth requirements.

E8C06 What is the necessary bandwidth of a 13-WPM international Morse code transmission?

A. Approximately 13 Hz
B. Approximately 26 Hz
C. Approximately 52 Hz
D. Approximately 104 Hz

C You can calculate the necessary bandwidth of a CW signal with this formula

$$BW = B \times K$$

where BW = the necessary bandwidth of the signal in Hz, B = the baud rate of the transmission; and K = a factor relating to shape of the keying envelope.

Divide WPM by 1.2 to convert to baud rate. K is typically between 3 (soft keying) and 5 (hard keying). A typical value for K is 4.8. The necessary bandwidth for a CW signal then becomes

$$BW = \frac{WPM}{1.2} \times K = \frac{13}{1.2} \times 4.8 = 52 \text{ Hz}$$

If you look at the formula you can see that with a K of 4.8, the necessary bandwidth of a CW signal in Hz is the speed in WPM times 4.

E8C07   What is the necessary bandwidth of a 170-hertz shift, 300-baud ASCII transmission?

    A. 0.1 Hz
    B. 0.3 kHz
    C. 0.5 kHz
    D. 1.0 kHz

C   An AFSK data signal is generated by injecting two audio tones, separated by the correct shift into the microphone input of an SSB transmitter. The necessary bandwidth for this type of data transmission is

$$BW = (K \times Shift) + B$$

where BW = the necessary bandwidth in hertz; K = a constant that for Amateur Radio you can assume to be 1.2; Shift = frequency shift in hertz; and B = data rate in bauds.

Plug the values into the formula and you get

$$BW = (K \times Shift) + B = (1.2 \times 170) + 300 = 504 \text{ Hz}$$

This is approximately 0.5 kHz.

E8C08   What is the necessary bandwidth of a 4800-Hz frequency shift, 9600-baud ASCII FM transmission?

    A. 15.36 kHz
    B. 9.6 kHz
    C. 4.8 kHz
    D. 5.76 kHz

A   Use the same formula as in the previous question. Substitute the given values and you get

$$BW = (K \times Shift) + B = (1.2 \times 4800) + 9600 = 15.36 \text{ kHz}$$

E8C09   What term describes a wide-bandwidth communications system in which the transmitted carrier frequency varies according to some predetermined sequence?

    A. Amplitude compandored single sideband
    B. AMTOR
    C. Time-domain frequency modulation
    D. Spread-spectrum communication

D   The keys here are the phrases "wide bandwidth" and "predetermined sequence." This describes spread spectrum (SS) communication.

E8C10 Which of these techniques causes a digital signal to appear as wide-band noise to a conventional receiver?

A. Spread-spectrum
B. Independent sideband
C. Regenerative detection
D. Exponential addition

A This is one of the reasons spread-spectrum techniques were invented; to prevent conventional receivers from receiving the signal. Because the energy of the SS signal is distributed in a pseudo-random fashion across its occupied bandwidth, it appears as noise to an ordinary receiver.

E8C11 What spread-spectrum communications technique alters the center frequency of a conventional carrier many times per second in accordance with a pseudo-random list of channels?

A. Frequency hopping
B. Direct sequence
C. Time-domain frequency modulation
D. Frequency compandored spread-spectrum

A Frequency hopping spread spectrum alters the center frequency of a conventional carrier many times per second in accordance with a pseudo-random list of channels.

E8C12 What spread-spectrum communications technique uses a high speed binary bit stream to shift the phase of an RF carrier?

A. Frequency hopping
B. Direct sequence
C. Binary phase-shift keying
D. Phase compandored spread-spectrum

B Direct sequence spread spectrum uses a very fast binary bit stream to shift the phase of an RF carrier.

E8C13 What makes spread-spectrum communications resistant to interference?

A. Interfering signals are removed by a frequency-agile crystal filter
B. Spread-spectrum transmitters use much higher power than conventional carrier-frequency transmitters
C. Spread-spectrum transmitters can hunt for the best carrier frequency to use within a given RF spectrum
D. Only signals using the correct spreading sequence are received

D One of the major advantages to spread-spectrum communications is that only signals using the correct spreading sequence are received. That makes this mode highly resistant to interference.

E8C14 What is the advantage of including a parity bit with an ASCII character stream?

A. Faster transmission rate
B. The signal can overpower interfering signals
C. Foreign language characters can be sent
D. Some types of errors can be detected

D   The parity bit in an ASCII character stream allows the receiving system to tell whether the data character contained an odd or even number of "1" bits. If the parity matches the received data, the character is likely to have been received correctly. Parity checking can detect "one-bit" errors, in which the state of a single bit is altered. It can not be relied upon to detect errors of more than one bit.

E8C15 What is one advantage of using JT-65 coding?

A. Uses only a 65 Hz bandwidth
B. Virtually perfect decoding of signals well below the noise
C. Easily copied by ear if necessary
D. Permits fast-scan TV transmissions over narrow bandwidth

B   JT-65 is an example of a protocol and modulation technique that uses repeated and predictable sequences of data to be combined in a way that diminishes the effect of noise. This is called "processing gain" and it allows signals to be demodulated with signal levels well below that of the noise.

E8D   Waves, measurements, and RF grounding: peak-to-peak values, polarization; RF grounding

E8D01 What is the easiest voltage amplitude parameter to measure when viewing a pure sine wave signal on an oscilloscope?

A. Peak-to-peak voltage
B. RMS voltage
C. Average voltage
D. DC voltage

A   Of the choices given, the peak-to-peak voltage, or the full span of the wave, is the easiest to measure on an oscilloscope. There is no need for calculations after you've taken the measurement.

E8D02 What is the relationship between the peak-to-peak voltage and the peak voltage amplitude of a symmetrical waveform?

A. 0.707:1
B. 2:1
C. 1.414:1
D. 4:1

B In a symmetrical waveform the positive peak voltage and the negative peak have equal amplitudes, but with opposite polarities. In other words, peak positive voltage equals peak negative voltage times –1. That means that peak-to-peak voltage is twice the peak voltage. The two are related by a 2:1 ratio. You can see this relationship in Figure E8D02.

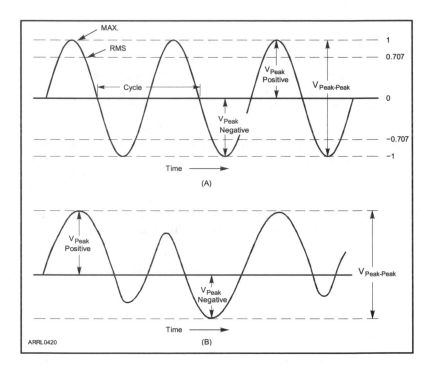

Figure E8D02 — You can measure the peak negative, peak positive and peak-to-peak values of a waveform on an oscilloscope display. (A) shows a sine waveform and (B) shows a complex waveform.

E8D03 What input-amplitude parameter is valuable in evaluating the signal-handling capability of a Class A amplifier?

A. Peak voltage
B. RMS voltage
C. Average power
D. Resting voltage

A Since Class A amplifiers are linear, the output is an amplified, but undistorted, representation of the input. You can evaluate a Class A amplifier by determining the greatest peak voltage that the amplifier can handle without distortion.

E8D04 What is the PEP output of a transmitter that has a maximum peak of 30 volts to a 50-ohm load as observed on an oscilloscope?

A. 4.5 watts
B. 9 watts
C. 16 watts
D. 18 watts

B The formula for average power is the square of the RMS voltage divided by the impedance

$$P_{AVG} = \frac{V_{RMS}^2}{Z}$$

You are given the peak voltage, so you'll have to convert that to RMS and then calculate the power at the peak, which is the peak envelope power (PEP). Use the peak-to-RMS formula

$$V_{RMS} = \frac{V_{Peak}}{\sqrt{2}}$$

Now use this conversion in the power formula, and you find

$$PEP = \left(\frac{V_{Peak}}{\sqrt{2}}\right)^2 \div Z = \frac{V_{Peak}^2}{2 \times Z} = \frac{30^2}{2 \times 50} = \frac{900}{100} = 9 \text{ W}$$

E8D05　If an RMS-reading AC voltmeter reads 65 volts on a sinusoidal waveform, what is the peak-to-peak voltage?

A. 46 volts
B. 92 volts
C. 130 volts
D. 184 volts

D　The peak voltage is computed by 1.414 (the square root of 2) times the RMS voltage and the peak-to-peak is twice that value. Substituting the given values,

$$V_{P-P} = 2 \times 1.414 \times 65 \text{ V} = 184 \text{ V}$$

E8D06　What is the advantage of using a peak-reading wattmeter to monitor the output of a SSB phone transmitter?

A. It is easier to determine the correct tuning of the output circuit
B. It gives a more accurate display of the PEP output when modulation is present
C. It makes it easier to detect high SWR on the feed-line
D. It can determine if any "flat-topping" is present during modulation peaks

B　A peak-reading wattmeter indicates the peak envelope power output (PEP) of a modulated signal, as long as the transmitter and amplifier are properly adjusted to eliminate spurious outputs.

E8D07　What is an electromagnetic wave?

A. Alternating currents in the core of an electromagnet
B. A wave consisting of two electric fields at right angles to each other
C. A wave consisting of an electric field and a magnetic field oscillating at right angles to each other
D. A wave consisting of two magnetic fields at right angles to each other

C　An electromagnetic wave derives its name from the fact that it has both an electric and a magnetic field component. The fields exist at right angles to one another. Which means, for example, when one is horizontal the other is vertical and vice-versa.

E8D08 Which of the following best describes electromagnetic waves traveling in free space?

A. Electric and magnetic fields become aligned as they travel
B. The energy propagates through a medium with a high refractive index
C. The waves are reflected by the ionosphere and return to their source
D. Changing electric and magnetic fields propagate the energy

D The electromagnetic wave is made up of an electric field and a magnetic field that are at right angles to each other. It is the changing electric and magnetic fields that propagate electromagnetic energy across the vacuum of free space. These fields are always at right angles as the wave travels along its path.

E8D09 What is meant by circularly polarized electromagnetic waves?

A. Waves with an electric field bent into a circular shape
B. Waves with a rotating electric field
C. Waves that circle the Earth
D. Waves produced by a loop antenna

B It is possible to generate waves whose electric and magnetic fields rotate, maintaining their right-angle orientation. This condition is called circular polarization. It is particularly helpful to use circular polarization in satellite communication, where polarization tends to shift.

E8D10 What is the polarization of an electromagnetic wave if its magnetic field is parallel to the surface of the Earth?

A. Circular
B. Horizontal
C. Elliptical
D. Vertical

D Polarization is defined in terms of the electric field, and the two fields are perpendicular to each other. If the magnetic field is parallel to the surface, then the electric field is perpendicular to the surface and the polarization is vertical.

E8D11 What is the polarization of an electromagnetic wave if its magnetic field is perpendicular to the surface of the Earth?

A. Horizontal
B. Circular
C. Elliptical
D. Vertical

A Again, polarization is defined in terms of the electric field and the two fields are perpendicular to each other. If the magnetic field is perpendicular to the surface then the electric field is horizontal and the polarization is horizontal.

E8D12 At approximately what speed do electromagnetic waves travel in free space?

A. 300 million meters per second
B. 186,300 meters per second
C. 186,300 feet per second
D. 300 million miles per second

A The speed of light is approximately 300,000,000 meters per second. In English customary units this is about 186,300 miles (not feet) per second. Be careful when you come to this question.

E8D13 What type of meter should be used to monitor the output signal of a voice-modulated single-sideband transmitter to ensure you do not exceed the maximum allowable power?

A. An SWR meter reading in the forward direction
B. A modulation meter
C. An average reading wattmeter
D. A peak-reading wattmeter

D If you wish to avoid violating the power rules, then you'll need to monitor your peak power (PEP) output. You can do that with a peak-reading wattmeter.

E8D14 What is the average power dissipated by a 50-ohm resistive load during one complete RF cycle having a peak voltage of 35 volts?

A. 12.2 watts
B. 9.9 watts
C. 24.5 watts
D. 16 watts

A You can calculate the correct answer plugging the appropriate values into the formula for average power. Remember that you'll have to convert peak voltage to RMS.

$$P_{AVG} = \left(\frac{V_{Peak}}{\sqrt{2}}\right)^2 \div R = \frac{V_{Peak}^{\;2}}{2 \times R} = \frac{35^2}{2 \times 50} = \frac{1225}{100} = 12.25\ W$$

E8D15 If an RMS reading voltmeter reads 34 volts on a sinusoidal waveform, what is the peak voltage?

A. 123 volts
B. 96 volts
C. 55 volts
D. 48 volts

D For a sinusoidal waveform, multiply the RMS voltage by 1.414 (square root of 2) to get the peak voltage. When you multiply 34 V_{RMS} by 1.414 you get 48 V_{Peak}.

E8D16 Which of the following is a typical value for the peak voltage at a common household electrical outlet?

A. 240 volts
B. 170 volts
C. 120 volts
D. 340 volts

B The common household voltage is 120 V, but this is the RMS value. The question is asking for the peak voltage. To convert this to a peak measurement, use this formula

$$V_{Peak} = V_{RMS} \times 1.414 = 120 \times 1.414 = 170\ V$$

The relationships among RMS, peak and peak-to-peak are shown in Figure E8D02.

E8D17 Which of the following is a typical value for the peak-to-peak voltage at a common household electrical outlet?

    A. 240 volts
    B. 120 volts
    C. 340 volts
    D. 170 volts

C The normal household RMS voltage is 120 V. To convert this to a peak-to-peak measurement, use this formula

$$V_{Pk\text{-}Pk} = 2 \times V_{Peak} = 2 \times V_{RMS} \times 1.414 = 2 \times 120 \times 1.414 = 340 \text{ V}$$

E8D18 Which of the following is a typical value for the RMS voltage at a common household electrical power outlet?

    A. 120-V AC
    B. 340-V AC
    C. 85-V AC
    D. 170-V AC

A By now you should be able to spot 120 V ac as the correct answer.

E8D19 What is the RMS value of a 340-volt peak-to-peak pure sine wave?

    A. 120-V AC
    B. 170-V AC
    C. 240-V AC
    D. 300-V AC

A This question requires you to solve a previous problem in reverse. To compute RMS from the peak-to-peak value of a sinusoid, you can start with the formula that you used earlier

$$V_{Pk\text{-}Pk} = 2 \times V_{Peak} = 2 \times V_{RMS} \times 1.414$$

Now, rearrange the formula and solve the problem.

$$V_{RMS} = \frac{V_{Pk-Pk}}{2 \times \sqrt{2}} = \frac{340}{2.828} = 120 \text{ V}$$

Antennas

There will be eight questions on your Extra class examination from the Antennas subelement. These eight questions will be taken from the eight groups of questions labeled E9A through E9H.

E9A Isotropic and gain antennas: definition; used as a standard for comparison; radiation pattern; basic antenna parameters: radiation resistance and reactance, gain, beamwidth, efficiency

E9A01 Which of the following describes an isotropic antenna?

A. A grounded antenna used to measure earth conductivity
B. A horizontal antenna used to compare Yagi antennas
C. A theoretical antenna used as a reference for antenna gain
D. A spacecraft antenna used to direct signals toward the earth

C You'll never see an isotropic radiator because it is a theoretical (mathematical) concept, useful only for comparing antenna performance. It is a point-source radiator located in space that exhibits no directivity in any direction. In other words, it radiates equally in all directions and its radiation pattern is perfectly spherical.

E9A02 How much gain does a ½-wavelength dipole have compared to an isotropic antenna?

A. 1.55 dB
B. 2.15 dB
C. 3.05 dB
D. 4.30 dB

B A ½ wavelength dipole has directivity. That means that there is more radiation in some directions than there is in others. Apart from the influence of the earth or other objects, a dipole has a 2.15 dB gain when compared to an isotropic radiator. (Some sources give 2.2 dB, rather than 2.15 dB.)

E9A03 Which of the following antennas has no gain in any direction?

A. Quarter-wave vertical
B. Yagi
C. Half-wave dipole
D. Isotropic antenna

D By definition, an isotropic radiator transmits uniformly in all directions. Gain is created by a departure from this uniform pattern.

E9A04 Why would one need to know the feed point impedance of an antenna?

A. To match impedances for maximum power transfer from a feed line
B. To measure the near-field radiation density from a transmitting antenna
C. To calculate the front-to-side ratio of the antenna
D. To calculate the front-to-back ratio of the antenna

A Causing as much of the available RF energy to be radiated as possible is accomplished, in part, by matching the antenna impedance to the feed line. To match the impedances, you must know the antenna's feed point impedance or create an adjustable matching system.

E9A05 Which of the following factors determine the radiation resistance of an antenna?

A. Transmission-line length and antenna height
B. Antenna height and conductor length/diameter ratio, and location of nearby conductive objects
C. It is a physical constant and is the same for all antennas
D. Sunspot activity and time of day

B An antenna's location with respect to nearby objects — especially the Earth — helps determine the radiation resistance. So does the conductors' length-to-diameter (L/D) ratio.

E9A06 What is the term for the ratio of the radiation resistance of an antenna to the total resistance of the system?

A. Effective radiated power
B. Radiation conversion loss
C. Antenna efficiency
D. Beamwidth

C   Radiation resistance is an assumed resistance that represents the power actually radiated from the antenna. Real, or ohmic, resistance in the system dissipates energy as heat. Total resistance is the sum of radiation resistance plus ohmic resistance. The ratio of the radiation resistance of an antenna to the total resistance of the system represents the antenna's efficiency, the percentage of total power radiated as electromagnetic waves.

E9A07 What is included in the total resistance of an antenna system?

A. Radiation resistance plus space impedance
B. Radiation resistance plus transmission resistance
C. Transmission-line resistance plus radiation resistance
D. Radiation resistance plus ohmic resistance

D   As explained in the previous question, the total resistance of an antenna system is the sum of the radiation resistance plus ohmic resistance.

E9A08 What is a folded dipole antenna?

A. A dipole one-quarter wavelength long
B. A type of ground-plane antenna
C. A dipole constructed from one wavelength of wire forming a very thin loop
D. A hypothetical antenna used in theoretical discussions to replace the radiation resistance

C A folded dipole antenna is a wire antenna that consists of one full wavelength of wire — unlike the one-half wavelength of wire in a regular dipole — in a very thin loop. It is fed at the center of one of the parallel wires as shown in Figure E9A08. The parallel wires act to raise the feed point impedance, which is useful in some applications.

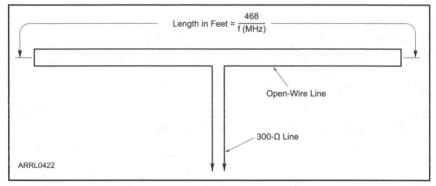

Figure E9A08 — A half-wave folded dipole consists of one full wavelength of wire in a very thin loop.

E9A09 What is meant by antenna gain?

A. The numerical ratio relating the radiated signal strength of an antenna in the direction of maximum radiation to that of a reference antenna
B. The numerical ratio of the signal in the forward direction to that in the opposite direction
C. The ratio of the amount of power radiated by an antenna compared to the transmitter output power
D. The final amplifier gain minus the transmission-line losses (including any phasing lines present)

A Gain measures the directivity of an antenna compared to a reference antenna. For example, the peak gain of a dipole is 2.15 dB greater than that of an isotropic antenna.

E9A10 What is meant by antenna bandwidth?

    A. Antenna length divided by the number of elements
    B. The frequency range over which an antenna satisfies a performance requirement
    C. The angle between the half-power radiation points
    D. The angle formed between two imaginary lines drawn through the element ends

B Bandwidth is the frequency range over which an antenna can be expected to meet some specified level of performance, such as an SWR value. For example, "SWR bandwidth" might refer to the range over which the antenna exhibits an SWR of 2:1 or less. Antennas also have gain bandwidth and front-to-back bandwidth.

E9A11 How is antenna efficiency calculated?

    A. (radiation resistance / transmission resistance) x 100%
    B. (radiation resistance / total resistance) x 100%
    C. (total resistance / radiation resistance) x 100%
    D. (effective radiated power / transmitter output) x 100%

B Calculate the efficiency of an antenna by dividing the radiation resistance by the total resistance. Multiply by 100 to get the answer in percent.
Total resistance is radiation resistance plus ohmic resistance. See also the discussion for E9A06.

E9A12 How can the efficiency of an HF quarter-wave grounded vertical antenna be improved?

    A. By installing a good radial system
    B. By isolating the coax shield from ground
    C. By shortening the vertical
    D. By reducing the diameter of the radiating element

A To significantly improve antenna efficiency, lower the antenna system's ohmic resistance. In the case of a typical HF grounded vertical antenna, losses in the ground system are the chief source of ohmic resistance. Installing a good ground radial system can lower this resistance.

E9A13 Which is the most important factor that determines ground losses for a ground-mounted vertical antenna operating in the 3-30 MHz range?

A. The standing-wave ratio
B. Base current
C. Soil conductivity
D. Base impedance

C Ground-mounted ¼ wavelength vertical antennas depend on a ground system to create an electrical image that creates the "missing half" of the ½ wavelength dipole. If the ground system consists largely of soil, losses are determined by the soil's conductivity. This is why installing a ground system of radial wires can greatly reduce ground system losses.

E9A14 How much gain does an antenna have over a ½-wavelength dipole when it has 6 dB gain over an isotropic antenna?

A. 3.85 dB
B. 6.0 dB
C. 8.15 dB
D. 2.79 dB

A A ½ wavelength dipole has a gain of 2.15 dB relative to an isotropic antenna. (Some sources give the figure as 2.2 dB.) The gain relative to the dipole would then be 6 dB – 2.15 dB = 3.85 dB.

E9A15 How much gain does an antenna have over a ½-wavelength dipole when it has 12 dB gain over an isotropic antenna?

A. 6.17 dB
B. 9.85 dB
C. 12.5 dB
D. 14.15 dB

B Again, the ½ wavelength dipole has a gain of 2.15 dB relative to an isotropic antenna. The gain relative to the dipole would then be 12 dB – 2.15 dB = 9.85 dB.

E9A16 What is meant by the radiation resistance of an antenna?

A. The combined losses of the antenna elements and feed line
B. The specific impedance of the antenna
C. The value of a resistance that would dissipate the same amount of power as that radiated from an antenna
D. The resistance in the atmosphere that an antenna must overcome to be able to radiate a signal

C Radiation resistance is the equivalent resistance that would dissipate the same amount of power as that radiated from an antenna.

E9B Antenna patterns: E and H plane patterns; gain as a function of pattern; antenna design (computer modeling of antennas); Yagi antennas

E9B01 What determines the free-space polarization of an antenna?

A. The orientation of its magnetic field (H Field)
B. The orientation of its free-space characteristic impedance
C. The orientation of its electric field (E Field)
D. Its elevation pattern

C The polarization of an antenna is determined by the orientation of its radiated electric field. For example, if the electric field is horizontally oriented relative to the Earth, we say that the polarization of the antenna is horizontal.

E9B02 In the antenna radiation pattern shown in Figure E9-1, what is the 3-dB beamwidth?

A. 75 degrees
B. 50 degrees
C. 25 degrees
D. 30 degrees

B By looking at the figure, the strength of the radiated signal drops 3 dB at about ±25 degrees from the peak of the main lobe. That makes the beamwidth about 50 degrees.

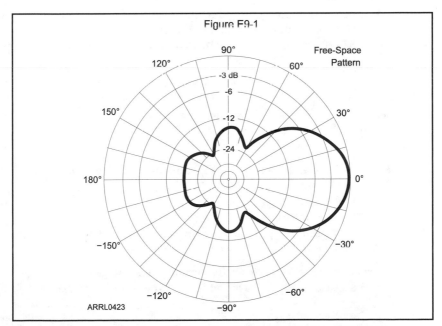

Figure E9-1 — Use this drawing for questions E9B02 through E9B04.

E9B03 In the antenna radiation pattern shown in Figure E9-1, what is the front-to-back ratio?

A. 36 dB
B. 18 dB
C. 24 dB
D. 14 dB

B The antenna gain at 0 degrees is 0 dB while the gain at 180 degrees is about halfway between −12 and −24 dB on this scale. You should read this as a ratio of 18 dB.

E9B04 In the antenna radiation pattern shown in Figure E9-1, what is the front-to-side ratio?

A. 12 dB
B. 14 dB
C. 18 dB
D. 24 dB

B The antenna gain at 0 degrees is 0 dB while the gain at 90 degrees is a bit less than −12 dB and can be estimated as a ratio of 14 dB.

E9B05 What may occur when a directional antenna is operated at different frequencies within the band for which it was designed?

A. Feed-point impedance may become negative
B. The E-field and H-field patterns may reverse
C. Element spacing limits could be exceeded
D. The gain may exhibit significant variations

D Antenna gain changes with frequency because the electrical length and spacing of its elements change. In antennas that are sensitive to small changes in element characteristics, such as Yagi antennas, gain and other parameters can change significantly across a frequency band.

E9B06 What usually occurs if a Yagi antenna is designed solely for maximum forward gain?

A. The front-to-back ratio increases
B. The front-to-back ratio decreases
C. The frequency response is widened over the whole frequency band
D. The SWR is reduced

B You can design a Yagi antenna for maximum forward gain, but the pattern ratios such as front-to-back and front-to-side ratio usually decrease.

E9B07   If the boom of a Yagi antenna is lengthened and the elements are properly retuned, what usually occurs?

A. The gain increases
B. The SWR decreases
C. The front-to-back ratio increases
D. The gain bandwidth decreases rapidly

A   As a rule, the gain of a Yagi antenna is directly proportional to the length of its boom. This requires that the number, spacing and lengths of the elements be properly adjusted. In other words, the gain of a Yagi increases as the boom is lengthened if the elements are properly retuned.

E9B08   How does the total amount of radiation emitted by a directional (gain) antenna compare with the total amount of radiation emitted from an isotropic antenna, assuming each is driven by the same amount of power?

A. The total amount of radiation from the directional antenna is increased by the gain of the antenna
B. The total amount of radiation from the directional antenna is stronger by its front to back ratio
C. There is no difference between the two antennas
D. The radiation from the isotropic antenna is 2.15 dB stronger than that from the directional antenna

C   The total amount of power radiated is the same for both antennas (ignoring heat losses from the real or ohmic resistance) but the directional antenna focuses it in one or more directions. Gain only means that the antenna has directivity, not that additional power is created.

E9B09   How can the approximate beamwidth of a directional antenna be determined?

A. Note the two points where the signal strength of the antenna is 3 dB less than maximum and compute the angular difference
B. Measure the ratio of the signal strengths of the radiated power lobes from the front and rear of the antenna
C. Draw two imaginary lines through the ends of the elements and measure the angle between the lines
D. Measure the ratio of the signal strengths of the radiated power lobes from the front and side of the antenna

A   The beamwidth is typically defined in terms of the half-power points. These are the points on either side and closest to the main lobe where the gain 3 dB below the peak gain. The beamwidth is the angular difference between these two points.

E9B10 What type of computer program technique is commonly used for modeling antennas?

A. Graphical analysis
B. Method of Moments
C. Mutual impedance analysis
D. Calculus differentiation with respect to physical properties

B Antenna modeling software uses the Method of Moments to analyze antenna performance. This analysis technique divides an antenna into segments, computes the current in each segment, and vector sums the radiation resulting from the currents in all segments.

E9B11 What is the principle of a Method of Moments analysis?

A. A wire is modeled as a series of segments, each having a distinct value of current
B. A wire is modeled as a single sine-wave current generator
C. A wire is modeled as a series of points, each having a distinct location in space
D. A wire is modeled as a series of segments, each having a distinct value of voltage across it

A The Method of Moments analysis technique divides wires into segments and assigns an appropriate current value to each segment. Each current value is called a "moment."

E9B12 What is a disadvantage of decreasing the number of wire segments in an antenna model below the guideline of 10 segments per half-wavelength?

A. Ground conductivity will not be accurately modeled
B. The resulting design will favor radiation of harmonic energy
C. The computed feed-point impedance may be incorrect
D. The antenna will become mechanically unstable

C By reducing the number of the segments so that they become longer, the assumption that current is constant across the segment becomes less valid. This degrades the accuracy of the final calculations.

E9B13 Which of the following is a disadvantage of NEC-based antenna modeling programs?

A. They can only be used for simple wire antennas
B. They are not capable of generating both vertical and horizontal polarization patterns
C. Computing time increases as the number of wire segments is increased
D. All of these answers are correct

C Computing time increases dramatically with antenna complexity. However, most available personal computers can perform NEC-based modeling without excessive computation time.

E9B14 What does the abbreviation NEC stand for when applied to antenna modeling programs?

A. Next Element Comparison
B. Numerical Electromagnetics Code
C. National Electrical Code
D. Numeric Electrical Computation

B The Numerical Electromagnetics Code was developed, validated and placed in the public domain. Many antenna programs make use of its fundamental algorithms. There are several versions, with the latest being NEC-4.

E9B15 What type of information can be obtained by submitting the details of a proposed new antenna to a modeling program?

A. SWR vs. frequency charts
B. Polar plots of the far-field elevation and azimuth patterns
C. Antenna gain
D. All of these answers are correct

D Antenna modeling programs can calculate all of these parameters and many more. The ability of individual amateurs to have sophisticated modeling programs has created many opportunities for amateur antenna experimentation.

E9C   Wire and phased vertical antennas: beverage antennas; terminated and resonant rhombic antennas; elevation above real ground; ground effects as related to polarization; take-off angles

E9C01   What is the radiation pattern of two ¼-wavelength vertical antennas spaced ½-wavelength apart and fed 180 degrees out of phase?

A. A cardioid
B. Omnidirectional
C. A figure-8 broadside to the axis of the array
D. A figure-8 oriented along the axis of the array

D   Phased-array antennas create their radiation patterns by controlling the phase of the feed point current and the spacing of the antennas. The resulting phase differences of the radiated signal from each element create the radiation pattern's lobes and nulls through constructive and destructive interference, respectively.

The radiation pattern of two ¼ wavelength vertical antennas spaced ½ wavelength apart and fed 180 degrees out of phase is a figure-8 end-fire in line with the antennas. See the Figure E9C01.

E9C02   What is the radiation pattern of two ¼-wavelength vertical antennas spaced ¼-wavelength apart and fed 90 degrees out of phase?

A. A cardioid
B. A figure-8 end-fire along the axis of the array
C. A figure-8 broadside to the axis of the array
D. Omnidirectional

A   In this case, the pattern is a cardioid, as you can see from Figure E9C01.

Figure E9C01 — Horizontal directive patterns of two phased verticals, spaced and phased as indicated. In these plots, you are looking directly onto the axis of the antennas (the antennas are perpendicular to the page) and the pair of antennas are arranged along a line running vertically ("North-South") through the plot. The uppermost (northern) element is lagging if there is a phase difference between the two antennas. The antennas are identical with the same magnitude of current in each. Gain is stated with respect to a single vertical.

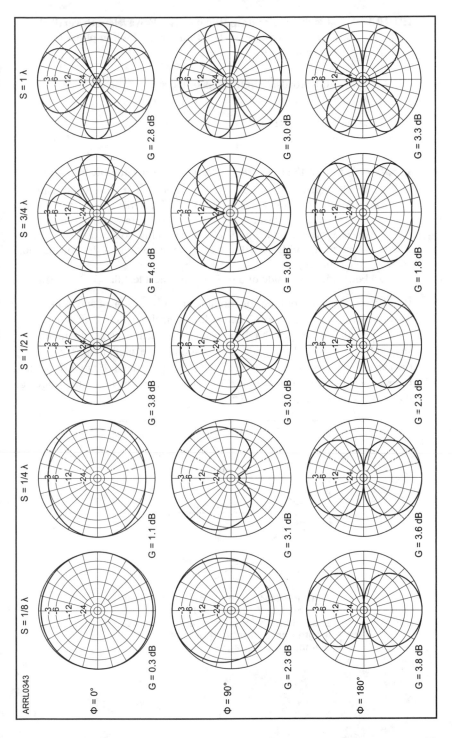

E9C03 What is the radiation pattern of two ¼-wavelength vertical antennas spaced ½-wavelength apart and fed in phase?

A. Omnidirectional
B. A cardioid
C. A Figure-8 broadside to the axis of the array
D. A Figure-8 end-fire along the axis of the array

C Refer to Figure E9C01, and you'll find that the pattern is a figure-8 broadside to the antennas.

E9C04 Which of the following describes a basic rhombic antenna?

A. Unidirectional; four-sided, each side one quarter-wavelength long; terminated in a resistance equal to its characteristic impedance
B. Bidirectional; four-sided, each side one or more wavelengths long; open at the end opposite the transmission line connection
C. Four-sided; an LC network at each corner except for the transmission connection;
D. Four-sided, each side of a different physical length

B There are two major types of rhombic antennas: resonant and non-resonant (or terminated). You'll need to know the differences between the two and the advantages and disadvantages of each. Both are diamond shaped (four-sided) and each side is at least one wavelength long. The resonant rhombic antenna is open at the end opposite the feed point. It has a bidirectional pattern. See Figure E9C04. The rhombic antenna is the source of the ARRL's diamond logo.

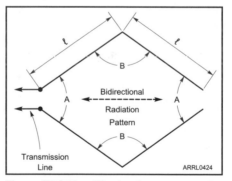

Figure E9C04 — The basic or resonant rhombic is a symmetric diamond-shaped antenna; all legs are the same length and opposite angles of the diamond are equal.

E9C05 **What are the main advantages of a terminated rhombic antenna?**

A. Wide frequency range, high gain and high front-to-back ratio
B. High front-to-back ratio, compact size and high gain
C. Unidirectional radiation pattern, high gain and compact size
D. Bidirectional radiation pattern, high gain and wide frequency range

A A terminated rhombic differs from a resonant rhombic in that the end opposite the feed point is terminated with a resistor. That termination changes the pattern from bidirectional to unidirectional and gives the antenna a wide frequency range (bandwidth). This is not a compact antenna.

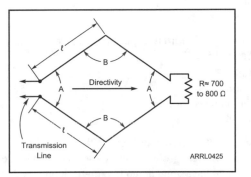

Figure E9C05 — The terminated or nonresonant rhombic has a terminating resistor added at the end opposite the feed point. The main effect of this resistor is to change the pattern from one that is primarily bidirectional to one that is primarily unidirectional.

E9C06 **What are the disadvantages of a terminated rhombic antenna for the HF bands?**

A. The antenna has a very narrow operating bandwidth
B. The antenna produces a circularly polarized signal
C. The antenna requires a large physical area and 4 separate supports
D. The antenna is more sensitive to man-made static than any other type

C It takes a large area to properly install an HF rhombic antenna. You'll also need four sturdy supports to hold up its four corners. This is true for all HF rhombics, terminated or not.

E9C07 **What is the effect of a terminating resistor on a rhombic antenna?**

A. It reflects the standing waves on the antenna elements back to the transmitter
B. It changes the radiation pattern from bidirectional to unidirectional
C. It changes the radiation pattern from horizontal to vertical polarization
D. It decreases the ground loss

B The main effect of a terminating resistor on a rhombic antenna is to change the radiation pattern from essentially bidirectional to essentially unidirectional.

E9C08 **What type of antenna pattern over real ground is shown in Figure E9-2?**

A. Elevation
B. Azimuth
C. Radiation resistance
D. Polarization

A The antenna pattern shown in Figure E9-2 is an elevation pattern. An elevation pattern over real ground usually shows only half a circle. Any radiation that would have gone down into the ground is reflected back into space above the Earth, and that energy is added to energy radiated directly from the antenna, creating the lobes and nulls of the elevation pattern.

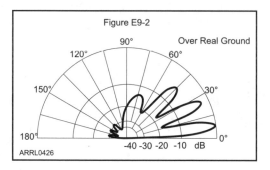

Figure E9-2

Figure E9-2 — Use this drawing for questions E9C08 through E9C11.

E9C09   What is the elevation angle of peak response in the antenna radiation pattern shown in Figure E9-2?

A. 45 degrees
B. 75 degrees
C. 7.5 degrees
D. 25 degrees

C   By looking at the radiation pattern, you can see the largest lobe is centered at 7.5 degrees above the horizon. This is the peak response.

E9C10   What is the front-to-back ratio of the radiation pattern shown in Figure E9-2?

A. 15 dB
B. 28 dB
C. 3 dB
D. 24 dB

B   The back lobes rise to just above the –30 dB circle and the main lobe reaches the 0-dB circle. This means you have a front-to-back ratio of about 28 dB.

E9C11   How many elevation lobes appear in the forward direction of the antenna radiation pattern shown in Figure E9-2?

A. 4
B. 3
C. 1
D. 7

A   The forward direction includes those lobes between 0 degrees and 90 degrees and there are four lobes in that range.

E9C12   How is the far-field elevation pattern of a vertically polarized antenna affected by being mounted over seawater versus rocky ground?

A. The low-angle radiation decreases
B. The high-angle radiation increases
C. Both the high- and low-angle radiation decrease
D. The low-angle radiation increases

D   Seawater has excellent conductivity and rocky ground is a poor conductor. If you mount a vertically polarized antenna over seawater, it will have the effect of increasing the low-angle radiation as compared to a similar antenna over rocky ground.

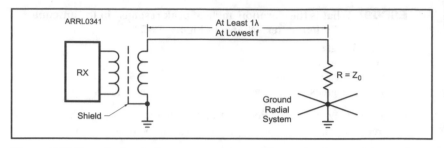

Figure E9C13 — A simple Beverage antenna with terminating resistor and matching transformer for the feed line to the receiver.

E9C13 When constructing a Beverage antenna, which of the following factors should be included in the design to achieve good performance at the desired frequency?

A. Its overall length must not exceed ¼ wavelength
B. It must be mounted more than 1 wavelength above ground
C. It should be configured as a four-sided loop
D. It should be one or more wavelengths long

D The Beverage antenna (invented by H.H. Beverage in 1922) works best when it is greater than one wavelength in length.

E9C14 How would the electric field be oriented for a Yagi with three elements mounted parallel to the ground?

A. Vertically
B. Horizontally
C. Right-hand elliptically
D. Left-hand elliptically

B This configuration, with the elements in a horizontal position, produces horizontal polarization. That is to say, the electric field is aligned horizontally.

E9C15 What strongly affects the shape of the far-field, low-angle elevation pattern of a vertically polarized antenna?

A. The conductivity and dielectric constant of the soil in the area of the antenna
B. The radiation resistance of the antenna and matching network
C. The SWR on the transmission line
D. The transmitter output power

A The radiation pattern of an antenna over real ground is always affected by the electrical conductivity and dielectric constant of the soil. This is especially true of the low-elevation-angle far-field pattern of a vertically polarized antenna. The low-angle radiation pattern from a vertically polarized antenna mounted over seawater will be much stronger than for a similar antenna mounted over rocky soil, for example. The less the loss in the ground around the antenna, the more radiation will occur at low vertical angles.

E9C16 This question has been removed by the Question Pool Committee.

E9C17 What is the main effect of placing a vertical antenna over an imperfect ground?

A. It causes increased SWR
B. It changes the impedance angle of the matching network
C. It reduces low-angle radiation
D. It reduces losses in the radiating portion of the antenna

C See the discussion for E9C15.

E9D **Directional antennas: gain; satellite antennas; antenna beamwidth; losses; SWR bandwidth; antenna efficiency; shortened and mobile antennas; grounding**

E9D01 How does the gain of a parabolic dish antenna change when the operating frequency is doubled?

A. Gain does not change
B. Gain is multiplied by 0.707
C. Gain increases 6 dB
D. Gain increases 3 dB

C For a parabolic dish, the gain is proportional to the square of the frequency so if you double the frequency, you will raise the gain by a factor of four. In dB units, a factor of 4 is a gain of 6 dB.

E9D02 What is one way to produce circular polarization when using linearly polarized antennas?

A. Stack two Yagis, fed 90 degrees out of phase, to form an array with the respective elements in parallel planes
B. Stack two Yagis, fed in phase, to form an array with the respective elements in parallel planes
C. Arrange two Yagis perpendicular to each other with the driven elements at the same point on the boom and fed 90 degrees out of phase
D. Arrange two Yagis collinear to each other, with the driven elements fed 180 degrees out of phase

C Two Yagi antennas built on the same boom, with elements placed perpendicular to each other, form the basis of a circularly polarized antenna. The driven elements are located at the same position along the boom, so they lie on the same plane, which is perpendicular to the boom. The driven elements are then fed 90 degrees out of phase. You can see how this works in Figure E9D02.

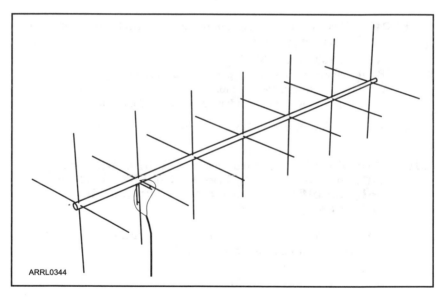

ARRL0344

Figure E9D02 — A circularly polarized antenna built from two Yagis on the same boom.

E9D03 How does the beamwidth of an antenna vary as the gain is increased?

A. It increases geometrically
B. It increases arithmetically
C. It is essentially unaffected
D. It decreases

D   The higher the gain, the more directive the antenna pattern. In other words, the beamwidth will decrease as gain increases.

E9D04 Why is it desirable for a ground-mounted satellite communications antenna system to be able to move in both azimuth and elevation?

A. In order to track the satellite as it orbits the earth
B. So the antenna can be pointed away from interfering signals
C. So the antenna can be positioned to cancel the effects of Faraday rotation
D. To rotate antenna polarization to match that of the satellite

A   For terrestrial communications, you would mount a beam antenna parallel to the Earth so that you can aim the antenna at the receiving station. A single rotator turns your antenna to any desired compass heading, or azimuth. As a satellite passes your location, however, it may be high above the horizon for a significant part of the time. You'll need a second rotator to change the antenna elevation angle to track the satellite across the sky much as you would need to change the bearing of a horizontal antenna to track a mobile station as it moves.

E9D05 For a shortened vertical antenna, where should a loading coil be placed to minimize losses and produce the most effective performance?

A. Near the center of the vertical radiator
B. As low as possible on the vertical radiator
C. As close to the transmitter as possible
D. At a voltage node

A   As you move a loading coil higher on the vertical, it requires more inductance to bring the antenna to resonance. More inductance results in more ohmic losses. However, as the coil moves up, the radiation resistance of the antenna goes up, which results in higher efficiency. You'll find that a loading coil placed near the center of the vertical balances the two effects for the most effective performance from the antenna.

E9D06 Why should an HF mobile antenna loading coil have a high ratio of reactance to resistance?

A. To swamp out harmonics
B. To maximize losses
C. To minimize losses
D. To minimize the Q

C   Higher resistance means lower efficiency. In other words, as the resistance goes up so do the ohmic and system losses. You want to use a loading coil with a high ratio of reactance to resistance, because that minimizes losses.

E9D07 What is a disadvantage of using a multiband trapped antenna?

A. It might radiate harmonics
B. It can only be used for single-band operation
C. It is too sharply directional at lower frequencies
D. It must be neutralized

A   Because the trap antenna is a multiband antenna, it can do a good job of radiating harmonics.

E9D08 What happens to the bandwidth of an antenna as it is shortened through the use of loading coils?

A. It is increased
B. It is decreased
C. No change occurs
D. It becomes flat

B   The bandwidth will decrease if loading coils are used to shorten an antenna.

E9D09 What is an advantage of using top loading in a shortened HF vertical antenna?

A. Lower Q
B. Greater structural strength
C. Higher losses
D. Improved radiation efficiency

D   Top loading is a technique that can reduce loading-coil losses. The method requires that a capacitive "hat" be added above the coil. A capacitive hat is so-named because it consists of wire spokes or rings that have a hat-like appearance. The added capacitance at the top of the antenna allows a smaller value of loading inductance. This results in lower system losses, thus improving antenna radiation efficiency.

E9D10   What is the approximate feed-point impedance at the center of a folded dipole antenna?

A. 300 ohms
B. 72 ohms
C. 50 ohms
D. 450 ohms

A   The input impedance of a dipole is approximately 73 ohms. In a folded dipole antenna, the current is divided between the two parallel wires. That means that the current at the feed point of a folded dipole is half of what it is in a dipole. From the power formula $P = I^2 \times Z$, you can see that with the same power and half the current, you'll also have four times the impedance — in this case approximately 300 ohms.

E9D11   Why is a loading coil often used with an HF mobile antenna?

A. To improve reception
B. To lower the losses
C. To lower the Q
D. To cancel capacitive reactance

D   An HF mobile antenna is usually less than ¼ wavelength long, which means the feed point has a capacitive reactance. The loading coil is inductive and is used to tune out the capacitive reactance so that the antenna is resonant.

E9D12   What is an advantage of using a trapped antenna?

A. It has high directivity in the higher-frequency bands
B. It has high gain
C. It minimizes harmonic radiation
D. It may be used for multi-band operation

D   A trap antenna may be used for multiband operation. See also E9D07.

E9D13   What happens at the base feed-point of a fixed-length HF mobile antenna as the frequency of operation is lowered?

A. The resistance decreases and the capacitive reactance decreases
B. The resistance decreases and the capacitive reactance increases
C. The resistance increases and the capacitive reactance decreases
D. The resistance increases and the capacitive reactance increases

B   Lowering the frequency of operation has the same effect as shortening the antenna. (The antenna becomes shorter electrically.) That means that the resistance decreases and the capacitive reactance increases.

E9D14 Which of the following types of conductor would be best for minimizing losses in a station's RF ground system?

A. A resistive wire, such as a spark-plug wire
B. A thin, flat copper strap several inches wide
C. A cable with 6 or 7 18-gauge conductors in parallel
D. A single 12 or 10 gauge stainless steel wire

B Ground conductors should have both low ohmic resistance and low inductance to minimize impedance at high frequencies. The type of conductor that best achieves those goals is a thin, flat metal strip, such as copper.

E9D15 Which of these choices would provide the best RF ground for your station?

A. A 50-ohm resistor connected to ground
B. A connection to a metal water pipe
C. A connection to 3 or 4 interconnected ground rods driven into the Earth
D. A connection to 3 or 4 interconnected ground rods via a series RF choke

C A safety ground may consist of a single ground rod driven deep into the ground, but at RF, it is more important to distribute the RF current widely in the ground near the surface. For that reason, several shallow ground rods are often used, connected together as a single ground.

E9E Matching: matching antennas to feed lines; power dividers

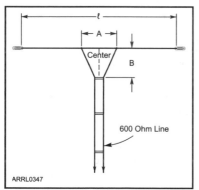

ARRL0347

Figure E9E01 — The delta matching system is used to match a high-impedance transmission line to a lower-impedance antenna. The feed line attaches to the driven element in two places, spaced a fraction of a wavelength on each side of the element center.

E9E01 What system matches a high-impedance transmission line to a lower impedance antenna by connecting the line to the driven element in two places spaced a fraction of a wavelength each side of element center?

A. The gamma matching system
B. The delta matching system
C. The omega matching system
D. The stub matching system

B A delta match network matches a high-impedance transmission line to a lower impedance antenna by connecting the line to the driven element in two places spaced a fraction of a wavelength each side of the element center. The principle is illustrated in Figure E9E01.

E9E02 What is the name of an antenna matching system that matches an unbalanced feed line to an antenna by feeding the driven element both at the center of the element and at a fraction of a wavelength to one side of center?

A. The gamma match
B. The delta match
C. The omega match
D. The stub match

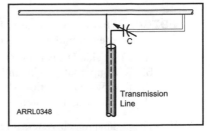

Figure E9E02 — The gamma matching system is used to match an unbalanced feed line to an antenna. The feed line attaches to the center of the driven element and to a point that is a fraction of a wavelength to one side of center.

A A gamma matching system matches an unbalanced feed line to an antenna by feeding the driven element both at the center of the element and at a fraction of a wavelength to one side of center. This is illustrated in Figure E9E02.

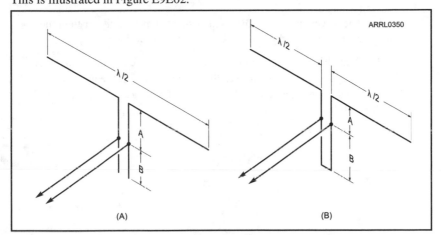

Figure E9E03 — The stub matching system uses a short perpendicular section of transmission line connected to the feed line near the antenna. Dimensions A + B equal ¼ wavelength.

E9E03 What is the name of the matching system that uses a short perpendicular section of transmission line connected to the feed line near the antenna?

A. The gamma match
B. The delta match
C. The omega match
D. The stub match

D The stub matching system uses a short perpendicular section of transmission line connected to the feed line near the antenna. You can see how this works in Figure E9E03.

E9E04 What is the purpose of the series capacitor in a gamma-type antenna matching network?

A. To provide DC isolation between the feed-line and the antenna
B. To compensate for the inductive reactance of the matching network
C. To provide a rejection notch to prevent the radiation of harmonics
D. To transform the antenna impedance to a higher value

B The gamma match system's feed line section parallel to the antenna element has some inductive reactance at the feed point. The adjustable capacitor cancels the reactance and brings the antenna system to resonance.

E9E05 How must the driven element in a 3-element Yagi be tuned to use a hairpin matching system?

A. The driven element reactance must be capacitive
B. The driven element reactance must be inductive
C. The driven element resonance must be lower than the operating frequency
D. The driven element radiation resistance must be higher than the characteristic impedance of the transmission line

A The Yagi driven element must be tuned so it has capacitive reactance. The hairpin inductance then cancels the capacitive reactance as it transforms the feed point impedance to a higher value that matches that of the feed line. See also E9E06.

E9E06 What is the equivalent lumped-constant network for a hairpin matching system on a 3-element Yagi?

A. Pi network
B. Pi-L network
C. L network
D. Parallel-resonant tank

C   The hairpin or beta match is an inductive network and the feed point is capacitive. Together they form the equivalent of an L network. This is illustrated Figure E9E06.

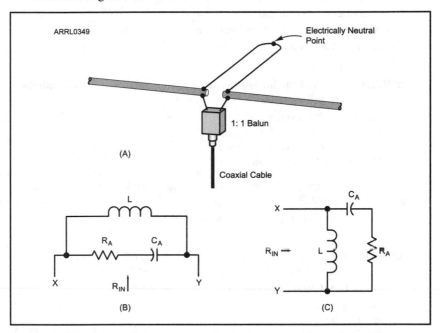

Figure E9E06 — The driven element of a Yagi antenna can be fed with a hairpin or beta matching system, as shown at A. Part B shows the lumped constant equivalent circuit, where R_A and C_A represent the antenna feed point impedance, and L represents the parallel inductance of the hairpin. Points X and Y represent the feed line connection. When the equivalent circuit is redrawn as shown at C, you can see that L and C_A form an L network to match the feed line to the antenna resistance R_A.

E9E07 What parameter best describes the interactions at the load end of a mismatched transmission line?

A. Characteristic impedance
B. Reflection coefficient
C. Velocity factor
D. Dielectric Constant

B The reflection coefficient is the ratio of reflected voltage (or current) to the incident voltage (or current) at the same point on a line. The reflection coefficient is determined by the relationship between the feed line characteristic impedance and the actual load impedance. That makes the reflection coefficient a good parameter to describe the interactions at the load end of a mismatched transmission line.

E9E08 Which of the following measurements describes a mismatched transmission line?

A. An SWR less than 1:1
B. A reflection coefficient greater than 1
C. A dielectric constant greater than 1
D. An SWR greater than 1:1

D If there is a mismatch, the SWR (Standing Wave Ratio) will be greater than 1:1. An SWR less than 1:1 is not possible. It is also not possible to have a dielectric constant or reflection coefficient greater than 1.

E9E09 Which of these matching systems is an effective method of connecting a 50-ohm coaxial cable feed-line to a grounded tower so it can be used as a vertical antenna?

A. Double-bazooka match
B. Hairpin match
C. Gamma match
D. All of these answers are correct

C A grounded tower can be thought of as one-half of a dipole driven element and a matching system used to match the base impedance of the tower to that of a feed line. The most convenient method of doing so is the gamma match.

E9E10 Which of these choices is an effective way to match an antenna with a 100-ohm terminal impedance to a 50-ohm coaxial cable feed-line?

A. Connect a ¼-wavelength open stub of 300-ohm twin-lead in parallel with the coaxial feed-line where it connects to the antenna
B. Insert a ½ wavelength piece of 300-ohm twin-lead in series between the antenna terminals and the 50-ohm feed cable
C. Insert a ¼-wavelength piece of 75-ohm coaxial cable transmission line in series between the antenna terminals and the 50-ohm feed cable
D. Connect ½ wavelength shorted stub of 75-ohm cable in parallel with the 50-ohm cable where it attaches to the antenna

C   A ¼ wavelength section of feed line with a characteristic impedance (Z_0) close to the geometric mean of the feed line impedance (Z_1) and antenna feed point impedance (Z_2), such that

$$Z_0 = \sqrt{Z_1 \, Z_2}$$

will match the two impedances. This is called a ¼ wave synchronous transformer.

E9E11 What is an effective way of matching a feed-line to a VHF or UHF antenna when the impedances of both the antenna and feed-line are unknown?

A. Use a 50-ohm 1:1 balun between the antenna and feed-line
B. Use the "universal stub" matching technique
C. Connect a series-resonant LC network across the antenna feed terminals
D. Connect a parallel-resonant LC network across the antenna feed terminals

B   The universal stub matching technique is illustrated in question E9E03.

E9E12 What is the primary purpose of a "phasing line" when used with an antenna having multiple driven elements?

A. It ensures that each driven element operates in concert with the others to create the desired antenna pattern
B. It prevents reflected power from traveling back down the feed-line and causing harmonic radiation from the transmitter
C. It allows single-band antennas to operate on other bands
D. It makes sure the antenna has a low-angle radiation pattern

A Phased-array antennas, such as those described by questions E9C01 through E9C03, depend on precise current and phase at the element feed points to create the desired patterns. Carefully constructed lengths of feed line are used so that the proper current and phasing relationships are created.

E9E13 What is the purpose of a "Wilkinson divider"?

A. It divides the operating frequency of a transmitter signal so it can be used on a lower frequency band
B. It is used to feed high-impedance antennas from a low-impedance source
C. It divides power equally among multiple loads while preventing changes in one load from disturbing power flow to the others
D. It is used to feed low-impedance loads from a high-impedance source

C The Wilkinson power divider is used to split the power from a single source into two or more equally divided portions for use in a phased array antenna system. It also helps isolate each output from each other and from the source.

E9F Transmission lines: characteristics of open and shorted feed lines: ⅛ wavelength; ¼ wave length; ½ wavelength; feed lines: coax versus open-wire; velocity factor; electrical length; transformation characteristics of line terminated in impedance not equal to characteristic impedance

E9F01 What is the velocity factor of a transmission line?

A. The ratio of the characteristic impedance of the line to the terminating impedance
B. The index of shielding for coaxial cable
C. The velocity of the wave in the transmission line multiplied by the velocity of light in a vacuum
D. The velocity of the wave in the transmission line divided by the velocity of light in a vacuum

D The velocity factor of a transmission line is the velocity of the wave in the line divided by the velocity of light in a vacuum.

E9F02   What determines the velocity factor in a transmission line?

A. The termination impedance
B. The line length
C. Dielectric materials used in the line
D. The center conductor resistivity

C   The presence of dielectric insulating materials reduces the velocity of an electromagnetic wave in a transmission line, since those waves travel more slowly in materials other than a vacuum.

E9F03   Why is the physical length of a coaxial cable transmission line shorter than its electrical length?

A. Skin effect is less pronounced in the coaxial cable
B. The characteristic impedance is higher in a parallel feed line
C. The surge impedance is higher in a parallel feed line
D. Electrical signals move more slowly in a coaxial cable than in air

D   The electrical length of a transmission line is measured in wavelengths at a given frequency. To calculate the physical length of a transmission line for a given electrical length, multiply the electrical (free space) length by the velocity factor. See questions E9F05, E9F06, and E9F09 for examples.

E9F04   What is the typical velocity factor for a coaxial cable with solid polyethylene dielectric?

A. 2.70
B. 0.66
C. 0.30
D. 0.10

B   The typical velocity factor for a coaxial cable with polyethylene dielectric is 0.66. In other words, the speed of an electromagnetic wave in typical RG-8 coax is about two-thirds the speed of light in a vacuum.

E9F05   What is the physical length of a coaxial transmission line that is electrically one-quarter wavelength long at 14.1 MHz? (Assume a velocity factor of 0.66.)

A. 20 meters
B. 2.3 meters
C. 3.5 meters
D. 0.2 meters

C   The 14.1 MHz frequency corresponds to a free-space wavelength of 21.3 meter (300 divided by the frequency in MHz). Since we wish to have a $\frac{1}{4}$ wavelength line, the free-space length would be 5.3 meters. This is multiplied by the velocity factor to give 3.5 meters.

E9F06 What is the physical length of a parallel conductor feed line that is electrically one-half wavelength long at 14.10 MHz? (Assume a velocity factor of 0.95.)

A. 15 meters
B. 20 meters
C. 10 meters
D. 71 meters

C Apply the same technique as in question E9F05. The free-space wavelength is 21.3 meters (300/14.1). We desire a ½ wavelength line, which would have a free-space length of 10.6 meters. This is multiplied by the velocity factor of .95 to give 10 meters.

E9F07 What characteristic will 450-ohm ladder line have at 50 MHz, as compared to 0.195-inch-diameter coaxial cable (such as RG-58)?

A. Lower loss
B. Higher SWR
C. Smaller reflection coefficient
D. Lower velocity factor

A The ladder line will typically have less than a tenth of the loss per 100 feet compared to RG-58.

E9F08 What is the term for the ratio of the actual speed at which a signal travels through a transmission line to the speed of light in a vacuum?

A. Velocity factor
B. Characteristic impedance
C. Surge impedance
D. Standing wave ratio

A The ratio of the actual velocity at which a signal travels through a transmission line to the speed of light in a vacuum is called the velocity factor.

E9F09 What would be the physical length of a typical coaxial transmission line that is electrically one-quarter wavelength long at 7.2 MHz? (Assume a velocity factor of 0.66)

A. 10 meters
B. 6.9 meters
C. 24 meters
D. 50 meters

B The free-space wavelength is 41.6 meters (300 divided by 7.2). You want a ¼ wavelength line, which would have a free-space length of 10.4 meters. Multiply this by the velocity factor (0.66) to give 6.9 meters.

E9F10 What kind of impedance does a ⅛-wavelength transmission line present to a generator when the line is shorted at the far end?

A. A capacitive reactance
B. The same as the characteristic impedance of the line
C. An inductive reactance
D. The same as the input impedance to the final generator stage

C This begins a series of questions about various line lengths and terminations. You'll find the answers in the Table E9-1, which shows input impedance to various length line sections that are terminated in a short or an open circuit. For example, from the table you see that a ⅛ wavelength transmission line that is shorted at the far end exhibits an inductive reactance at its input.

Table E9-1

Properties of Open and Shorted Feed-Line Sections

Length	Termination	Impedance
⅛ wavelength	Shorted	Inductive
⅛ wavelength	Open	Capacitive
¼ wavelength	Shorted	Very high impedance
¼ wavelength	Open	Very low impedance
½ wavelength	Shorted	Very low impedance
½ wavelength	Open	Very high impedance

E9F11 What kind of impedance does a ⅛-wavelength transmission line present to a generator when the line is open at the far end?

A. The same as the characteristic impedance of the line
B. An inductive reactance
C. A capacitive reactance
D. The same as the input impedance of the final generator stage

C From Table E9-1, the open-ended ⅛ wavelength transmission line looks capacitive to the generator.

E9F12 What kind of impedance does a ¼-wavelength transmission line present to a generator when the line is open at the far end?

A. A very high impedance
B. A very low impedance
C. The same as the characteristic impedance of the line
D. The same as the input impedance to the final generator stage

B From Table E9-1, the open-ended ¼ wavelength transmission line has a very low impedance at its input.

E9F13 What kind of impedance does a ¼-wavelength transmission line present to a generator when the line is shorted at the far end?

A. A very high impedance
B. A very low impedance
C. The same as the characteristic impedance of the transmission line
D. The same as the generator output impedance

A From Table E9-1, the shorted ¼ wavelength transmission line presents a very high impedance to the generator.

E9F14 What kind of impedance does a ½-wavelength transmission line present to a generator when the line is shorted at the far end?

A. A very high impedance
B. A very low impedance
C. The same as the characteristic impedance of the line
D. The same as the output impedance of the generator

B From Table E9-1, the shorted ½ wavelength transmission line presents a very low impedance at its input. It is useful to remember that impedance repeats every ½ wavelength along a transmission line.

E9F15   What kind of impedance does a ½-wavelength transmission line present to a generator when the line is open at the far end?

A. A very high impedance
B. A very low impedance
C. The same as the characteristic impedance of the line
D. The same as the output impedance of the generator

A   From Table E9-1, the open ½ wavelength transmission line presents a very high impedance to the generator.

E9F16   What is the primary difference between foam-dielectric coaxial cable as opposed to solid-dielectric cable, assuming all other parameters are the same?

A. Reduced safe operating voltage limits
B. Reduced losses per unit of length
C. Higher velocity factor
D. All of these answers are correct

D   The foam dielectric replaces some of the polyethylene with air, which has lower losses and increases the velocity factor. The tradeoff is that air is not quite as good an insulator as polyethylene and so the safe operating voltage limit is lower.

E9G   The Smith chart

E9G01   Which of the following can be calculated using a Smith chart?

A. Impedance along transmission lines
B. Radiation resistance
C. Antenna radiation pattern
D. Radio propagation

A   The Smith chart was developed for the purpose of graphically calculating impedance along a transmission line. It is unique in that ability.

E9G02 What type of coordinate system is used in a Smith chart?

    A. Voltage circles and current arcs
    B. Resistance circles and reactance arcs
    C. Voltage lines and current chords
    D. Resistance lines and reactance chords

B   The mnemonic for this question is R & R: resistance and reactance, circles and arcs. Impedance is resistance and reactance. These are plotted on a coordinate system that is shown in Figure E9G01. For simplicity, resistance and reactance coordinate systems are shown separately.

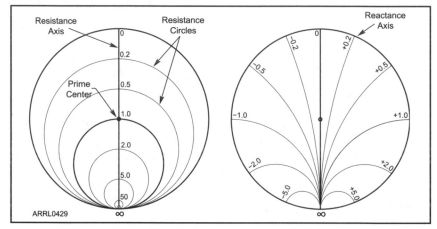

Figure E9G01 — A Smith Chart has resistance circles and reactance arcs. To help you visualize them, they are shown separately in this drawing.

E9G03 Which of the following is often determined using a Smith chart?

    A. Beam headings and radiation patterns
    B. Satellite azimuth and elevation bearings
    C. Impedance and SWR values in transmission lines
    D. Trigonometric functions

C   A Smith chart is not only used to plot impedances, it can also be used to calculate changes in resistance and reactance along a length of transmission line.

E9G04 What are the two families of circles and arcs that make up a Smith chart?

    A. Resistance and voltage
    B. Reactance and voltage
    C. Resistance and reactance
    D. Voltage and impedance

C   Since the Smith chart graphs impedances, the correct answer is resistance and reactance. Do you remember R & R?

E9G05 What type of chart is shown in Figure E9-3?

  A. Smith chart
  B. Free-space radiation directivity chart
  C. Elevation angle radiation pattern chart
  D. Azimuth angle radiation pattern chart

A The chart in Figure E9-3 is a Smith chart.

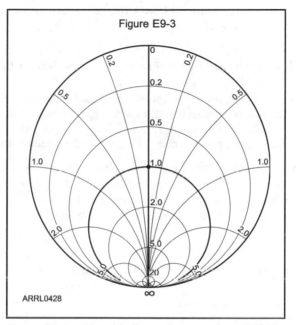

Figure E9-3

ARRL0428

Figure E9-3 — Use this figure for questions E9G05 through E9G07.

E9G06 On the Smith chart shown in Figure E9-3, what is the name for the large outer circle on which the reactance arcs terminate?

  A. Prime axis
  B. Reactance axis
  C. Impedance axis
  D. Polar axis

B If you noticed that the earlier questions always dealt with resistance and reactance, then you should be able to spot reactance axis as the right choice.

E9G07 On the Smith chart shown in Figure E9-3, what is the only straight line shown?

A. The reactance axis
B. The current axis
C. The voltage axis
D. The resistance axis

D That straight line is the resistance axis. The previous question asked for the reactance axis. It should not have surprised you to find that this question asks for the resistance axis.

E9G08 What is the process of normalization with regard to a Smith chart?

A. Reassigning resistance values with regard to the reactance axis
B. Reassigning reactance values with regard to the resistance axis
C. Reassigning impedance values with regard to the prime center
D. Reassigning prime center with regard to the reactance axis

C The process of assigning resistance values with regard to the value at prime center is called normalizing. To normalize values for a 50-ohm system, divide the resistance by 50. To convert from the chart values back to actual values, multiply by 50. Normalization permits the Smith chart to be used for any impedance value.

E9G09 What third family of circles is often added to a Smith chart during the process of solving problems?

A. Standing-wave ratio circles
B. Antenna-length circles
C. Coaxial-length circles
D. Radiation-pattern circles

A The correct answer is standing wave ratio circles. Smith chart plots can be used to give a measure of impedance mismatch. All of the points on the chart that result in the same SWR form a circle centered on the prime center.

E9G10 What do the arcs on a Smith chart represent?

A. Frequency
B. SWR
C. Points with constant resistance
D. Points with constant reactance

D When extended beyond the bounds of the Smith chart, the curved lines form reactance circles.

E9G11 How are the wavelength scales on a Smith chart calibrated?

A. In fractions of transmission line electrical frequency
B. In fractions of transmission line electrical wavelength
C. In fractions of antenna electrical wavelength
D. In fractions of antenna electrical frequency

B As you should remember, the Smith chart was developed for the purpose of graphically calculating impedance along a transmission line. For that reason, the scale is calibrated in terms of wavelength in a transmission line.

E9H Effective radiated power; system gains and losses; radio direction finding antennas

E9H01 What is the effective radiated power of a repeater station with 150 watts transmitter power output, 2-dB feed line loss, 2.2-dB duplexer loss and 7-dBd antenna gain?

A. 1977 watts
B. 78.7 watts
C. 420 watts
D. 286 watts

D This is the first of a series of questions related to dB and effective radiated power (ERP).

$$ERP = TPO \times \log^{-1}\left(\frac{\text{system loss or gain}}{10}\right)$$

where TPO = transmitter power output and (system loss or gain) = the sum of gains and losses starting at the transmitter and including antenna gain.

$$ERP = 150 \times \log^{-1}\left(\frac{-2 - 2.2 + 7}{10}\right) = 150 \times \log^{-1}(0.28) = 286 \text{ W}$$

E9H02   What is the effective radiated power of a repeater station with 200 watts transmitter power output, 4-dB feed line loss, 3.2-dB duplexer loss, 0.8-dB circulator loss and 10-dBd antenna gain?

A. 317 watts
B. 2000 watts
C. 126 watts
D. 300 watts

A   Use the same formula as before

$$ERP = 200 \times \log^{-1}\left(\frac{-4 - 3.2 - 0.8 + 10}{10}\right) = 200 \times \log^{-1}(0.2) = 317 \text{ W}$$

E9H03   What is the effective radiated power of a repeater station with 200 watts transmitter power output, 2-dB feed line loss, 2.8-dB duplexer loss, 1.2-dB circulator loss and 7-dBd antenna gain?

A. 159 watts
B. 252 watts
C. 632 watts
D. 63.2 watts

B   Use the same formula as before

$$ERP = 200 \times \log^{-1}\left(\frac{-2 - 2.8 - 1.2 + 7}{10}\right) = 200 \times \log^{-1}(0.1) = 252 \text{ W}$$

E9H04   What term describes station output (including the transmitter, antenna and everything in between), when considering transmitter power and system gains and losses?

A. Power factor
B. Half-power bandwidth
C. Effective radiated power
D. Apparent power

C   The intent of using effective radiated power (ERP) is to be able to compare different stations based on their transmitter output power and an equivalent antenna, the dipole. To do so, all of the antenna system gains and losses are combined into one compensating factor.

E9H05 What is the main drawback of a wire-loop antenna for direction finding?

A. It has a bidirectional pattern
B. It is non-rotatable
C. It receives equally well in all directions
D. It is practical for use only on VHF bands

A A wire-loop antenna has a bidirectional pattern. That means that it will not pinpoint the exact direction from which a signal is arriving.

E9H06 What is the triangulation method of direction finding?

A. The geometric angle of sky waves from the source are used to determine its position
B. A fixed receiving station plots three headings from the signal source on a map
C. Antenna headings from several different receiving stations are used to locate the signal source
D. A fixed receiving station uses three different antennas to plot the location of the signal source

C To perform triangulation, combine azimuth measurements from several diverse physical locations to determine the transmitter location. These azimuths or bearings are typically drawn on a map, and where they cross is the location of the transmitter.

E9H07 Why is an RF attenuator desirable in a receiver used for direction finding?

A. It narrows the bandwidth of the received signal
B. It eliminates the effects of isotropic radiation
C. It reduces loss of received signals caused by antenna pattern nulls
D. It prevents receiver overload from extremely strong signals

D If you think about it, the receiver in the "fox hunt" will need to receive signals with a large power range. Since the receiver may have a limited reception signal range, an attenuator makes a convenient helper in extending the dynamic range of the receiver by attenuating signals from a nearby fox.

E9H08 What is the function of a sense antenna?

    A. It modifies the pattern of a DF antenna array to provide a null in one direction

    B. It increases the sensitivity of a DF antenna array

    C. It allows DF antennas to receive signals at different vertical angles

    D. It provides diversity reception that cancels multipath signals

A   A sense antenna (sometimes called a sensing element) is a vertical antenna that is added to a loop antenna, and together they produce a cardioid reception pattern. This pattern is useful for direction finding.

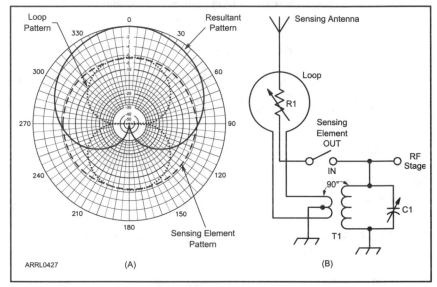

Figure E9H08 — At (A) the radiation pattern of a receiving loop antenna with sensing element. At (B) is a circuit for combining the signals from the two elements. C1 is adjusted for resonance at the operating frequency with T1.

E9H09 What is a receiving loop antenna?

    A. A large circularly-polarized antenna

    B. A small coil of wire tightly wound around a toroidal ferrite core

    C. One or more turns of wire wound in the shape of a large open coil

    D. Any antenna coupled to a feed line through an inductive loop of wire

C   A receiving loop antenna consists of one or more turns of wire wound in the shape of a large open coil. Receiving loops are very short in terms of wavelengths while transmitting loops are generally approximately one wavelength or more in circumference.

E9H10 How can the output voltage of a receiving loop antenna be increased?

A. By reducing the permeability of the loop shield
B. By increasing the number of wire turns in the loop and reducing the area of the loop structure
C. By reducing either the number of wire turns in the loop or the area of the loop structure
D. By increasing either the number of wire turns in the loop or the area of the loop structure

D   The strength of the signal coming from a loop antenna is proportional to the cross-sectional area of the antenna and the number of turns — the voltage increases as either of these parameters increase.

E9H11 Why is an antenna with a cardioid pattern desirable for a direction-finding system?

A. The broad-side responses of the cardioid pattern can be aimed at the desired station
B. The response characteristics of the cardioid pattern can assist in determining the direction of the desired station
C. The extra side lobes in the cardioid pattern can pinpoint the direction of the desired station
D. The high-radiation angle of the cardioid pattern is useful for short-distance direction finding

B   An antenna with either a sharp peak or a sharp null is good for direction finding because the peak or null can be used to "point" to the RF source. In the cardioid pattern, there is a deep, narrow null in one direction.

E9H12 What is an advantage of using a shielded loop antenna for direction finding?

A. It automatically cancels ignition noise pickup in mobile installations
B. It is electro-statically balanced against ground, giving better nulls
C. It eliminates tracking errors caused by strong out-of-band signals
D. It allows stations to communicate without giving away their position

B   A properly constructed shielded loop antenna is balanced against ground and so does not pick up stray signals that cause its pattern nulls to be shallower and less symmetrical. Both of those qualities are important in direction finding because they increase the precision and accuracy of the bearings to the signal source.

Safety

There will be one question on your Extra class examination from the Safety subelement. This question will be taken from the group of questions labeled E0A.

E0A Safety: amateur radio safety practices; RF radiation hazards; hazardous materials

E0A01 What, if any, are the differences between the radiation produced by radioactive materials and the electromagnetic energy radiated by an antenna?

A. There is no significant difference between the two types of radiation
B. Only radiation produced by radioactivity can injure human beings
C. RF radiation does not have sufficient energy to break apart atoms and molecules; radiation from radioactive sources does
D. Radiation from an antenna will damage unexposed photographic film, ordinary radioactive materials do not cause this problem

C Radiation from radioactive sources is called ionizing radiation because it has sufficient energy to create ions by knocking electrons away from their host atom. Wavelengths corresponding to ultraviolet light and shorter (X-rays, gamma rays, etc) are considered ionizing radiation. The very long wavelength radiation associated with radio is far too weak to ionize atoms and is considered non-ionizing. The only harmful effect of radio frequency radiation is heating at very high power levels.

E0A02 When evaluating exposure levels from your station at a neighbor's home, what must you do?

A. Make sure signals from your station are less than the controlled MPE limits
B. Make sure signals from your station are less than the uncontrolled MPE limits
C. Nothing; you need only evaluate exposure levels on your own property
D. Advise your neighbors of the results of your tests

B Because you do not control when your neighbor may be at home, the exposure is considered uncontrolled. When performing the required evaluation of your station, the uncontrolled MPE (maximum permissible exposure) limits are the ones you should use.

E0A03 Which of the following would be a practical way to estimate whether the RF fields produced by an amateur radio station are within permissible MPE limits?

A. Use a calibrated antenna analyzer
B. Use a hand calculator plus Smith-chart equations to calculate the fields
C. Walk around under the antennas with a neon-lamp probe to find the strongest fields
D. Use a computer-based antenna modeling program to calculate field strength at accessible locations

D It is easiest by far to use the approved modeling technique to determine if RF exposure from your station is close to or exceeds the MPE limits in accessible locations. Should the modeling technique indicate excessive exposure, following up with on-site measurements could be warranted.

E0A04 When evaluating a site with multiple transmitters operating at the same time, the operators and licensees of which transmitters are responsible for mitigating over-exposure situations?

A. Only the most powerful transmitter
B. Only commercial transmitters
C. Each transmitter that produces 5% or more of its maximum permissible exposure limit at accessible locations
D. Each transmitter operating with a duty-cycle greater than 50%

C In multitransmitter sites, such as hilltop facilities where there may be commercial facilities along with amateur repeaters, precisely determining responsibility for RF levels is impractical. Therefore, the rule was written so that responsibility is shared among the operators of all significant RF sources.

E0A05 What is one of the potential hazards of using microwaves in the amateur radio bands?

A. Microwaves are ionizing radiation
B. The high gain antennas commonly used can result in high exposure levels
C. Microwaves often travel long distances by ionospheric reflection
D. The extremely high frequency energy can damage the joints of antenna structures

B Because the wavelength of microwaves is a few centimeters or less, it is relatively straightforward to construct antennas with gain far in excess of that on longer wavelength bands. This can cause very high field strengths in the main lobe of high gain antennas such as dishes. Take special care when operating such antennas near people, for example during portable operation.

E0A06 Why are there separate electric (E) and magnetic (H) field MPE limits?

A. The body reacts to electromagnetic radiation from both the E and H fields
B. Ground reflections and scattering make the field impedance vary with location
C. E field and H field radiation intensity peaks can occur at different locations
D. All of these answers are correct

D In the far field of antennas, exposure from the E and H fields can be combined into a composite power density. Near the antenna or near reflecting surfaces, the E and H field intensity can vary significantly from the far field values. The ratio of E to H field intensity — the media impedance — can also be altered by scattering and reflection. In such cases, it may be prudent to evaluate exposure based on both the E and H field values.

E0A07 What is the "far-field" zone of an antenna?

A. The area of the ionosphere where radiated power is not refracted
B. The area where radiated power dissipates over a specified time period
C. The area where radiated field strengths are obstructed by objects of reflection
D. The area where the shape of the antenna pattern is independent of distance

D The far field begins where the maximum dimensions of the antenna become insignificant with respect to the distance from the antenna. Beyond this point — and there is no precise boundary — the shape of the antenna pattern stays constant with distance from the antenna. The level of the transmitted signal may change, but the shape of the radiation pattern stays the same.

E0A08 What does SAR measure?

A. Synthetic Aperture Ratio of the human body
B. Signal Amplification Rating
C. The rate at which RF energy is absorbed by the body
D. The rate of RF energy reflected from stationary terrain

C Specific Absorption Rate (SAR) measures the rate at which power from an electromagnetic field is absorbed by the human body. SAR changes with frequency and the shape of the body part.

E0A09 Which insulating material commonly used as a thermal conductor for some types of electronic devices is extremely toxic if broken or crushed and the particles are accidentally inhaled?

A. Mica
B. Zinc oxide
C. Beryllium Oxide
D. Uranium Hexaflouride

C Beryllium Oxide (BeO) is a white ceramic insulator that has the unusual property of also being an excellent thermal conductor. Most insulators are poor conductors of heat. So BeO is used in electronic devices as an insulating layer between the semiconductor itself and the metal case and in some vacuum tubes. BeO is not commonly encountered, but if there is any possibility of exposure, use protective gloves and take care not to touch or breath in any BeO dust. Check the manufacturer's data sheet to see if BeO is used in specific devices.

E0A10 What material found in some electronic components such as high-voltage capacitors and transformers is considered toxic?

A. Polychlorinated biphenyls
B. Polyethylene
C. Polytetrafluroethylene
D. Polymorphic silicon

A Polychlorinated biphenyls (PCBs) were once commonly added to insulating oils to improve their stability and insulating qualities. It was not discovered until later that they are carcinogenic (increase the probability of developing cancer) and so commercial production and use halted. Amateurs may encounter PCBs in the insulating oil of old oil-filled capacitors or dummy loads. To dispose of these components safely, consult your local electric utility for information.

E0A11 Which of these items might be a significant hazard when operating a klystron or cavity magnetron transmitter?

A. Hearing loss caused by high voltage corona discharge
B. Blood clotting from the intense magnetic field
C. Injury from radiation leaks that exceed the MPE limits
D. Ingestion of ozone gas from the cooling system

C Commercial equipment using klystrons or magnetrons (including microwave ovens) is carefully shielded to avoid microwave leakage. Homemade or converted equipment may not have the same extensive shielding and so should be operated with caution. The primary hazard is heating as the body absorbs the microwave energy.

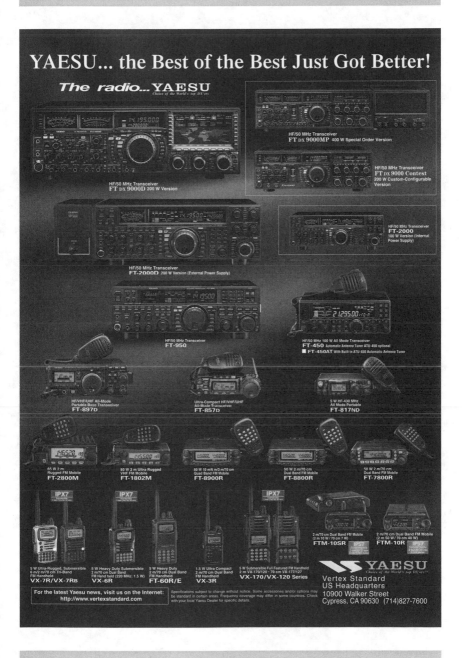

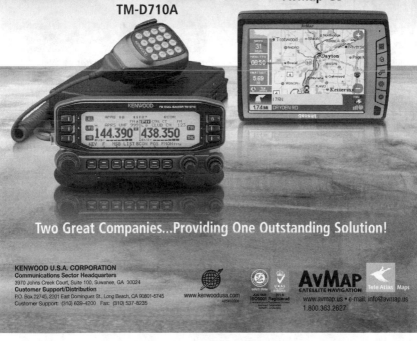

12 STORE BUYING POWER

HAM RADIO OUTLET

WORLDWIDE DISTRIBUTION

MAKE THE MOST OF YOUR EXTRA CLASS LICENSE!

IC-7800 All Mode Transceiver

• 160-6M @ 200W • Four 32 bit IF-DSPs+ 24 bit AD/DA converters • Two completely independent receivers • +40dBm 3rd order intercept point • Now with 3rd roofing filter

IC-7700 The Contester's Rig

• HF + 6m operation • 200W output power full duty cycle • +40dBm ultra high intercept point • IF DSP, user defined filters • Digital voice recorder

IC-756PROIII All Mode Transceiver

• 160-6M • 100W • Adjustable SSB TX bandwidth • Digital voice recorder • Auto antenna tuner • RX: 30 kHz to 60 MHz • Quiet, triple-conversion receiver • 32 bit IF-DSP • Low IMD roofing filter

IC-746PRO All Mode 160M-2M

• 160-2M* @ 100W • 32 bit IF-DSP+ 24 bit AD/DA converter • Selectable IF filter shapes for SSB & CW • Enhanced Rx performance

IC-718 HF Transceiver

• 160-10M* @ 100W • 12V Operation • Simple to Use • CW Keyer Built-in • One Touch Band Switching • Direct frequency input • VOX Built-in • Band stacking register • IF shift • 101 memories

IC-7000 All Mode Transceiver

• For the love of ham radio! • 160 - 10M, 6M @ 100W (40W AM), 2M @ 50W (20W AM), 70CM @ 35W (14W AM), all continually adjustable • 2x DSP • Digital IF filters • Digital voice recorder • 2.5" color TFT display • 503 memory channels • Remote control mic

IC-PW1 All Band Amplifier

• All band frequency coverage • 1KW amplifier • 100% duty cycle • Compact body • Detachable controller • Automatic antenna tuner

AT-180 Automatic Antenna Tuner

• Coax fed antenna system • 100W • Compact packaging, matches IC-706MKIIG/7000 • HF + 6m

IC-706MKIIG All Mode Transceiver

• Proven Performance • 160-10M*/6M/2M/70CM • All mode w/DSP • HF/6M @ 100W, 2M @ 50W, 440 MHz @ 20W • CTCSS encode/decode w/tone scan • Auto repeater • 107 alphanumeric memories

AH-4 Automatic Antenna Tuner

• Long wire, whip antenna systems • Mobile, waterproof packaging • 100W • Continuous tune 80m - 6m

*Except 60M Band. **Frequency coverage may vary. Refer to owner's manual for exact specs. © 2008 Icom America Inc. The Icom logo is a registered trademark of Icom Inc. 50012

ICOM

CALL TOLL FREE
Phone Hours: **Store Hours:**
9:30 AM -   10:00 AM - 5:30 PM
5:30 PM   Closed Sun

For Fax, incl. Hawaii, Alaska, Canada, call Router to nearest store; or HRO 800-Item can direct you. If the first line you call is busy, you may call another.

West...........800-854-6046
Mountain......800-444-9476
Southeast.....800-444-7927
Mid-Atlantic..800-444-4799
Northeast.....800-644-4476
New England..800-444-0047

Look for the
HRO Home Page
on the
World Wide Web
http://www.hamradio.com

AZ, CA, CO, GA, VA residents add sales tax. Prices, specifications, descriptions, subject to change without notice.

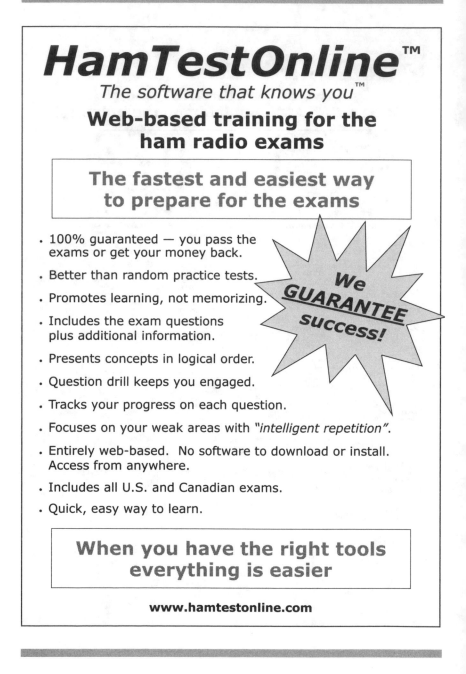

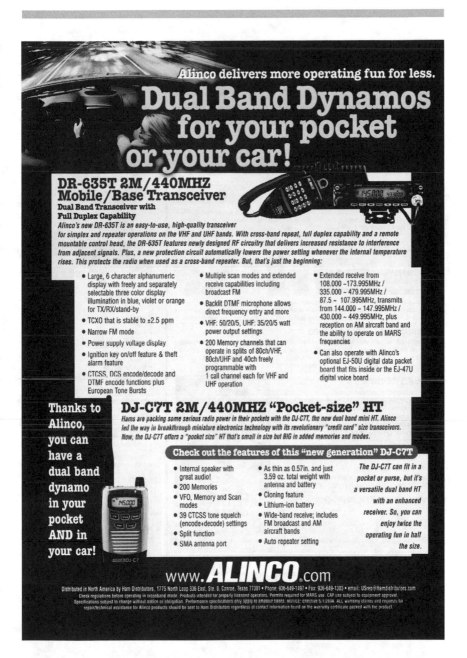

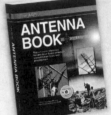

About the ARRL _____

The seed for Amateur Radio was planted in the 1890s, when Guglielmo Marconi began his experiments in wireless telegraphy. Soon he was joined by dozens, then hundreds, of others who were enthusiastic about sending and receiving messages through the air—some with a commercial interest, but others solely out of a love for this new communications medium. The United States government began licensing Amateur Radio operators in 1912.

By 1914, there were thousands of Amateur Radio operators—hams—in the United States. Hiram Percy Maxim, a leading Hartford, Connecticut inventor and industrialist, saw the need for an organization to band together this fledgling group of radio experimenters. In May 1914 he founded the American Radio Relay League (ARRL) to meet that need.

Today ARRL, with approximately 150,000 members, is the largest organization of radio amateurs in the United States. The ARRL is a not-for-profit organization that:
- promotes interest in Amateur Radio communications and experimentation
- represents US radio amateurs in legislative matters, and
- maintains fraternalism and a high standard of conduct among Amateur Radio operators.

At ARRL headquarters in the Hartford suburb of Newington, the staff helps serve the needs of members. ARRL is also International Secretariat for the International Amateur Radio Union, which is made up of similar societies in 150 countries around the world.

ARRL publishes the monthly journal *QST*, as well as newsletters and many publications covering all aspects of Amateur Radio. Its headquarters station, W1AW, transmits bulletins of interest to radio amateurs and Morse code practice sessions. The ARRL also coordinates an extensive field organization, which includes volunteers who provide technical information and other support services for radio amateurs as well as communications for public-service activities. In addition, ARRL represents US amateurs with the Federal Communications Commission and other government agencies in the US and abroad.

Membership in ARRL means much more than receiving *QST* each month. In addition to the services already described, ARRL offers membership services on a personal level, such as the ARRL Volunteer Examiner Coordinator Program and a QSL bureau.

Full ARRL membership (available only to licensed radio amateurs) gives you a voice in how the affairs of the organization are governed. ARRL policy is set by a Board of Directors (one from each of 15 Divisions) elected by the full members they represent. The day-to-day operation of ARRL HQ is managed by a Chief Executive Officer and his staff.

No matter what aspect of Amateur Radio attracts you, ARRL membership is relevant and important. There would be no Amateur Radio as we know it today were it not for the ARRL. We would be happy to welcome you as a member! (An Amateur Radio license is not required for Associate Membership.) For more information about ARRL and answers to any questions you may have about Amateur Radio, write or call:

ARRL—The national association for Amateur Radio
225 Main Street
Newington CT 06111-1494

Voice: 860-594-0200
Fax: 860-594-0259

E-mail: **hq@arrl.org**
Internet: **www.arrl.org/**

Prospective new amateurs call (toll-free):
800-32-NEW HAM (800-326-3942)
You can also contact us via e-mail at **newham@arrl.org**
or check out **ARRLWeb** at **www.arrl.org/**

"Join ARRL and experience the BEST of Ham Radio!"

ARRL Membership Benefits and Services:
- *QST* magazine — your monthly source of news, easy-to-read product reviews, and features for new hams!
- Technical Information Service — access to problem-solving experts!
- Members-only Web services — find information fast, anytime!
- ARRL clubs, mentors and volunteers — ready to welcome YOU!

FREE Book Offer!

I want to join ARRL.
Send me the FREE book I have selected (choose one)

- ☐ The ARRL Repeater Directory
- ☐ More Wire Antenna Classics
- ☐ The ARRL Emergency Communication Handbook

Name _____ Call Sign _____

Street _____

City _____ State _____ ZIP _____

Please check the appropriate one-year[1] rate:
- ☐ **$39 in US.**
- ☐ **Age 21 or younger rate, $20 in US** (see note*).
- ☐ **Canada $49.**
- ☐ **Elsewhere $62.**
- **Please indicate date of birth** _____ *

[1] 1-year membership dues include $15 for a 1-year subscription to QST. International 1-year rates include a $10 surcharge for surface delivery to Canada and a $23 surcharge for air delivery to other countries. Other US membership options available: Blind, Life, and QST by First Class postage. Contact ARRL for details.
*Age 21 or younger rate applies only if you are the oldest licensed amateur in your household.
International membership is available with an annual CD-ROM option (no monthly receipt of QST). Contact ARRL for details.
Dues subject to change without notice.

Sign up my family members, residing at the same address, as ARRL members too! They'll each pay only $8 for a year's membership, have access to ARRL benefits and services (except QST) and also receive a membership card.

☐ Sign up _____ family members @ $8 each = $ _____ . Attach their names & call signs (if any).

☐ Total amount enclosed, payable to ARRL $ _____ . (US funds drawn on a bank in the US.)

☐ Enclosed is $ _____ ($1.00 minimum) as a donation to the Legal Research and Resource Fund.

☐ Charge to: ☐ VISA ☐ MasterCard ☐ Amex ☐ Discover

Card Number _____ Expiration Date _____

Cardholder's Signature _____

Call Toll-Free (US) **1-888-277-5289**
Join Online **www.arrl.org/join** or
Clip and send to:

ARRL *The national association for* **AMATEUR RADIO**
225 Main Street
Newington, CT 06111-1494 USA

☐ If you do not want your name and address made available for non-ARRL related mailings, please check here.

ECLM QA-08

Please use this form to give us your comments on this book and what you'd like to see in future editions, or e-mail us at **pubsfdbk@arrl.org** (publications feedback). If you use e-mail, please include your name, call, e-mail address and the book title, edition and printing in the body of your message. Also indicate whether or not you are an ARRL member.

Please check the box that best answers these questions:
How well did this book prepare you for your exam?
□ Very Well □ Fairly Well □ Not Very Well
Did you pass? □ Yes □ No
Where did you purchase this book?
□ From ARRL directly □ From an ARRL dealer

Is there a dealer who carries ARRL publications within:
□ 5 miles □ 15 miles □ 30 miles of your location? □ Not sure.

If licensed, what is your license class? ─────────────────────

Name _____ ARRL member? □ Yes □ No

_____ Call Sign _____

Address _____
City, State/Province, ZIP/Postal Code _____
Daytime Phone ( ) _____ Age _____ E-mail _____
If licensed, how long? _____

Other hobbies _____

_____

Occupation _____

For ARRL use only	E Q&A
Edition	2 3 4 5 6 7 8 9 10 11
Printing	2 3 4 5 6 7 8 9 10 11

From _____

_____

_____

EDITOR, THE ARRL'S AMATEUR EXTRA Q&A
ARRL—THE NATIONAL ASSOCIATION FOR AMATEUR RADIO
225 MAIN STREET
NEWINGTON CT 06111-1494

--please fold and tape--